AF388972

Energy Performance in Buildings and Quality of Life

Energy Performance in Buildings and Quality of Life

Editor

Kristian Fabbri

MDPI • Basel • Beijing • Wuhan • Barcelona • Belgrade • Manchester • Tokyo • Cluj • Tianjin

Editor
Kristian Fabbri
University of Bologna
Italy

Editorial Office
MDPI
St. Alban-Anlage 66
4052 Basel, Switzerland

This is a reprint of articles from the Special Issue published online in the open access journal *Energies* (ISSN 1996-1073) (available at: https://www.mdpi.com/journal/energies/special_issues/building_energy_performance).

For citation purposes, cite each article independently as indicated on the article page online and as indicated below:

LastName, A.A.; LastName, B.B.; LastName, C.C. Article Title. *Journal Name* **Year**, *Article Number*, Page Range.

ISBN 978-3-03936-656-9 (Hbk)
ISBN 978-3-03936-657-6 (PDF)

Contents

About the Editor

Kristian Fabbri is an architect and consultant in building energy performance, energy services, indoor environmental quality. Fabbri has served as Adjunct Professor at the University of Bologna Department of Architecture since 2002, teaching courses on environmental physics techniques; building physics; building energy performance simulation; and energy behavior, comfort, and heritage buildings. He has qualified for Associate Professor to National Scientific Qualification Sector 09/C2 Environmental Physics and Nuclear Engineering ASN 2013 and 2018, sector 08/C1 Design and Technology Design of Architecture, 08/E2 Restoration and History of Architecture ASN 2016. Fabbri has been a consultant of Emilia-Romagna Region (Public Bodies), the public administration techno-legislative sector, SME trade associations, and professional training organizations since 2006. His research interest includes human behavior, indoor and outdoor comfort, energy poverty, heritage building, and building energy performance. Fabbri has published numerous research articles in international journals, conferences, and books. K. Fabbri Indoor thermal comfort perception (2013), M.Pretelli K.Fabbri Historic Indoor Microclimate of the Heritage Buildings (2018), S.Piraccini K.Fabbri Building a Passive House (2018); K. Fabbri Urban Fuel Poverty (2019). Outside of his research, Fabbri writes poetry (Triceratopo, Sauromachia) and dramas (Speer Architecture and Power)

Preface to "Energy Performance in Buildings and Quality of Life"

Buildings allow several kinds of human activity: work, eat, sleep, play, etc., and they have a role in determining quality of life: ugly and uncomfortable buildings can be the worst place to live. Building physics and building energy performance (BEP) research focuses on a several building-related issues: building typologies (e.g., schools, dwellings, social housing, heritage, etc.); construction technologies (e.g., wall, roof, windows including in prefabricated buildings, wall buildings, etc.); energy consumption and cost; energy saving; energy monitoring; and, furthermore, architectural design and the impact of specific building techniques, including material life cycle assessment and the role of materials on BEP. In this book, we highlight the relationship between BEP and quality of life.

We adopt the phrase "quality of life" because as all-inclusive term covering everything that impacts on habitual life, especially in terms of comfort, thermal comfort, and IAQ and household smartness (smart building, smart monitoring, or smart metering, following UE Directive 844/2018), as well as the reduction of energy poverty and the impact of buildings on the environment and global warming. Although the above list provides examples of specific topics of interest, our aim is, more broadly, to collect papers discussing the role of BEP in improving quality of life.

Kristian Fabbri
Editor

Article

Climate Change Effect on Building Performance: A Case Study in New York

Kristian Fabbri [1,*], **Jacopo Gaspari** [1] and **Licia Felicioni** [2]

[1] Department of Architecture, University of Bologna, Sede di Cesena, 47521 Cesena (FC), Italy; jacopo.gaspari@unibo.it

[2] University Centre for Energy Efficient Buildings, Czech Technical University, 27343 Prague, Buštěhrad, Czechia Republic; licia.felicioni@cvut.cz

* Correspondence: kristian.fabbri@unibo.it

Received: 18 May 2020; Accepted: 16 June 2020; Published: 18 June 2020

Abstract: The evidences of the influence of climate change (CC) in most of the key sectors of human activities are frequently reported by the news and media with increasing concern. The building sector, and particularly energy use in the residential sector, represents a crucial field of investigation as demonstrated by specific scientific literature. The paper reports a study on building energy consumption and the related effect on indoor thermal comfort considering the impacts of the Intergovernmental Panel on Climate Change (IPCC) 2018 report about temperature increase projection. The research includes a case study in New York City, assuming three different scenarios. The outcomes evidence a decrease in energy demand for heating and an increase in energy demand for cooling, with a relevant shift due to the summer period temperature variations. The challenge of the last decades for sustainable design was to increase insulation for improving thermal behavior, highly reducing the energy demand during winter time, however, the projections over the next decades suggest that the summer regime will represent a future and major challenge in order to reduce overheating and ensure comfortable (or at least acceptable) living conditions inside buildings. The growing request of energy for cooling is generating increasing pressure on the supply system with peaks in the case of extreme events that lead to the grid collapse and to massive blackouts in several cities. This is usually tackled by strengthening the energy infrastructure, however, the users' behavior and lifestyle will strongly influence the system capacity in stress conditions. This study focuses on the understanding of these phenomena and particularly on the relevance of the users' perception of indoor comfort, assuming the IPCC projections as the basis for a future scenario.

Keywords: climate change; building energy performance; IPCC; thermal comfort; building energy consumption; +1.5 degree; cooling increase

---

1. Introduction Concerning the Impacts of Climate Change

Since 1970, a significant increase in the global mean temperature and atmospheric carbon concentration has been registered, however, the relation with climate change was assumed as a global challenge quite recently. Despite many evidences of the interdependency of phenomena, the conventional approach to cope with climate change impacts is often locally tailored rather than considered in the global dimension. Climate change is much less relevant to the human condition than warming in cities [1] and it can strongly influence both energy production/distribution and demand in the built environment [2] while increasing the risks that extreme events can heavily affect power infrastructures [3]. Adaptation and mitigation actions are therefore to be considered within a broader perspective that often goes beyond the local borders [4].

The Paris Agreement (December 2015) [5] represents indeed a step forward for the 1997 Kyoto Protocol and the quite limited commitment coming from some major countries. The massive improvement of renewable energy solutions (RES), required to achieve the defined ambitious goals, is clearly linked to the energy demand for heating and cooling with relation to indoor and outdoor comfort conditions which are clearly affected by climate change impacts [6]. In order to ensure acceptable comfort levels [7], a number of protocols, standards, and regulations have been developed with relation to the building sector. IPCC's 2018 reports about climate change (CC) and buildings can actually be considered one of the most authoritative documents to properly approach the topic. Additionally, some specific considerations are required to properly discuss the progress of European countries and the United States among the major nations involved in the climate change challenge in the West world.

1.1. The 2018 IPCC's Report

The periodic assessment of climate science by the Intergovernmental Panel on Climate Change (IPCC) since the early 1990s is an essential contribution to the achievement of a global ontology on climate change [8]. 2018 IPCC's report—"Global warming of 1.5 °C"—is a key document within this challenge, offering a projection of the impact that a global warming of 1.5 °C and the related global Greenhouse Gas emission (GHG emission) routes would produce compared to pre-industrial levels. The report highlights several impacts that could be avoided by limiting global warming to 1.5 °C instead of 2 °C or more (such as sea level rise on a global scale by 2100 within a limit of 10 cm, the decrease of coral reefs of 70–90% instead of 99%, etc.). With the purpose of strengthening the global response to the threat of climate change, limiting global warming to 1.5 °C requires rapid changes in so many aspects of society contributing to sustainable development and to eradicate poverty [9]. Two major issues are related to the built environment: the relation between the building and the city scale as well as their performances with reference to the energy demand.

1.1.1. IPCC and the Role of Buildings and Cities

IPCC establishes different approaches to consider climate change as a collective action problem, and among them, it turned into a story of global mean temperature foreclosing the range of possible response options. By translating the multi-layered problem-complex of climate change (climates multiple) into a unitary global problem (climate singular), IPCC paved the way for a single policy agenda: emissions control monitored through a UN coordinated policy regime [10]. Through the adoption of the 2015 Paris Agreement, the climate regime is today replaced by a more decentralized and isolated climate governance order. By inviting states to propose nationally-appropriate mitigation and adaptation responses, the Paris Agreement has distributed responsibility for climate action across multiple actors, arenas, and sites [11]. However, few methods are actually available to analyze implications of climate change on building energy use [12]. The understanding of the combination of different factors within the design and operation stages is still crucial in contributing to improve the performance level.

1.1.2. Building Energy Performance (BEP)

The building sector contributes up to 40% of global energy consumption [13,14]. Despite the efforts spent in the last decades to reduce emissions, the level of GHG in the different scenarios projections demonstrates a dramatic rise in the coming future [15]. In this context, buildings represent a critical piece of a low-carbon future and their response to climate change solicitations has to be reliably predicted in order to take effective strategic design decisions in the mid-term. The role of building simulation tools has increased in its importance, giving the chance to speed up the design process, increase efficiency, and compare a broader range of design options. Building performance simulation (BPS) has been a suitable tool for helping to reach solutions for better energy efficiency [16]. Simulation provides a better understanding of the consequences of design decisions, increasing the effectiveness of the whole system [17] and it is now becoming increasingly relevant in post-construction phases of the building life-cycle (BLC), such as commissioning and operational management and control. Software simulations allow to predict the system behavior within unobserved conditions, allowing analysts to simultaneously consider the impacts on the overall performance [18]. During the simulations, many different strategies are evaluated to obtain the higher performance for a set of objectives (e.g., zero energy balance) [19]. The energy performance of the buildings is really influenced by heat transmission, thermal mass, solar heat gain through windows; thus, any action in these fields can be of help in reducing building energy consumption [20].

1.2. European Approach

According to the IPCC Fifth Assessment Report (AR5), 32% of global primary energy was spent by buildings, producing 19% of global emissions in 2010. Currently, these trends are still growing, reducing the chances to meet the targets agreed in the 2020 Climate and Energy Package [21], extended to 2030 [22] and 2050 within a long-term strategy [23,24]. The main efforts spent at the European Union (EU) level to reduce energy demand are addressed to decrease the needs for heating while ensuring adequate comfort conditions [25]. This led the EU to introduce the Energy Performance of Buildings Directive (EPBD Recast II [26] and EPBD Recast III [27]) and the Directive 2012/27/EU [28] on measures to help the EU to reach its 20% energy efficiency target by 2020 [29]. Each Member State can acknowledge the directives according to the most suitable and effective conditions within its own context in order to meet the targets (nearly Zero Energy Building (nZEB)) and to become more resilient to future climate conditions [30]. However, the effects of climate change and the evolving weather variables are influencing this process, impacting both the local network development [31] and energy use. Following IPCC projections, the energy demand for heating decreases due to the temperature variations and to the building performance improvement, while the demand for cooling is growing due to the increase of cities mean temperature.

1.3. USA Approach and New York Greening Actions

The recent withdrawal of the United States from the Paris Agreement requires a reflection on the situation of one of the biggest contributors to global emissions, which is also a country severely affected by the often dramatic consequences of extreme events [32]. Commercial and residential buildings account for about 40% of the primary energy demand in the US and respectively 9.9% and 5.4% of global GHG emissions.

Despite the decision at the federal level, some states, such as New York and California, continue their actions against the climate change autonomously. In September 2018, Senate Bill 100 (SB100) required the Renewable Portfolio Standard (RPS) for electric utilities from 50% to 60% by 2030 and further targeted 100% clean energy in all sectors by 2045 [33]. New York City has developed several scientific reports and local regulations for improving urban resilience to cope with climate change [34]. In 2018, the New York government recognized the urgency for renovating the energy network by providing micro-grids, to help the systems surviving natural disasters or energy demand peaks that recently brought to the grid collapse. Renewable energy is certainly part of these important environmental goals and the New York State energy plan of 2015 aims to shift from 11% [35] to 40% of energy needs by 2030. The "Reforming the Energy Vision" initiative [36,37] integrates technical, management, and marketing approaches to explore the compatibility of small-scale renewable technology with existing residential buildings, commercial, or public areas, serving local needs while remaining connected to the wider network system.

Projections of Annual Climate Change in New York

After the dramatic events of Hurricane Sandy, the New York City Panel on Climate Change (NPCC [38]) developed a system to collect and update information on the climate risk for the city with the purpose to both address the recovery after the event and to improve the resiliency.

The climate change impacts could be both short-term (storms and floods) and long-term (gradual changes in outdoor air temperature and sea level rise). As reported in NPPC [38], the long term impacts will be extremely likely events in the state of New York. Average annual temperatures are expected to increase up to 1.5 °C (3 °F) warmer by the 2020s, up to 3.3 °C (6 °F) warmer by the 2050s, and up to 5.5 °C (10 °F) warmer by the 2080s [38]. By 2100, the growing season could be about a month longer, with intense summers (extreme heat and heat waves) and milder winters. At the same time, rainfall will also increase noticeably, 1–8% by 2020, 3–12% by 2050, and 4–15% by 2080. These changes are associated with an increase in greenhouse gases and global warming [39]. The frequencies of heat and frost waves, intense rainfall, droughts, and coastal floods in the seven regions of the state will be subject to change in the coming decades, as predicted by the global climate model. These gradual changes will bring several consequences on built environment, and its energy consumption. Due to the huge difference between the outdoor air temperature and indoor temperature, the cooling systems will be responsible for the potential increase of energy end-use consumption. This is the state of the art so far following the IPCC report scenarios.

For investigating the impact of climate change on building energy consumption, the use of building simulation software, together with a set of forecasted weather data, is essential [40]. Within this general context, the methodology framework reported in this paper focused on the effects of a renovation project involving an iconic 14-story residential building in Red Hook, Brooklyn, New York considering the IPCC climate future projections of +1.5 °C overheating. The impact of climate change on indoor thermal comfort in the residential sector under different scenarios was also evaluated and reported in this paper.

2. Scope of the Research

The paper reports the outcomes of research aimed to evaluate the implications of IPCC climate change projections in the construction sector, applied to a case study in New York City.

The scope is to translate the outcomes of the research activities into concrete and practical indications that might be of help in replicating and driving the design process to tackle the climate change issue. Energy demand for heating and cooling is analyzed considering three weather scenarios (1958, 2017, 2100) according to IPCC projections and tested on a case study in Brooklyn, New York. In this way, simulations can take into account the predicted evolution of climate change [41]. The proposed study is primarily referred to the residential sector, which is the most relevant share in the building stock and one of the key targets of many regulations and governmental measures but can easily be adapted to other typologies. The paper aims also to provide a reflection on the effects of climate change to the thermal comfort, and the resistance of clothing (expressed in clo, m^2K/W) is used as a variable. The clothing resistance characteristics are expressed as an average value, i.e., they refer to all the clothing worn and not just some (e.g., only the shirt, etc.).

3. Research Methodology

The adopted methodology includes the calculation of Building Energy Performance Simulation (BEPS) for the case study considering three climatic scenarios. Weather data for the selected years are provided by Energyplus for 1958 and 2017—construction time and current condition respectively—while 2100 follows the IPCC temperature increase projection.

The use of weather data in simulation software like Energyplus, Ies.Ve, etc., can come from different sources (e.g., satellite data, weather stations, climate models, etc.) whose level of accuracy and reliability of the data source may change.

For the purposes of our research objectives, which is the evaluation of the variation over time of the simulation results, we considered the use of the data in the software database for the years 1958 and 2017. This choice may be questionable, but in this case, their accuracy supports the reliability of the adopted methodology. The approach combining simulations and scenarios is quite consolidated in the scientific literature, however, the focus on the effects of a temperature increase of +1.5 °C and the related overheating represent a novelty.

The research methodology is described following three sections:

1.   BEPS of the state of the art (it provides the starting conditions of the case study);
2.   Thermal comfort and clothes insulation;
3.   Thermo-economic renovation scenarios of improving building energy performance.

Figure 1 provides a flowchart of the research methodology (that is specifically designed to facilitate the replication to other case studies).

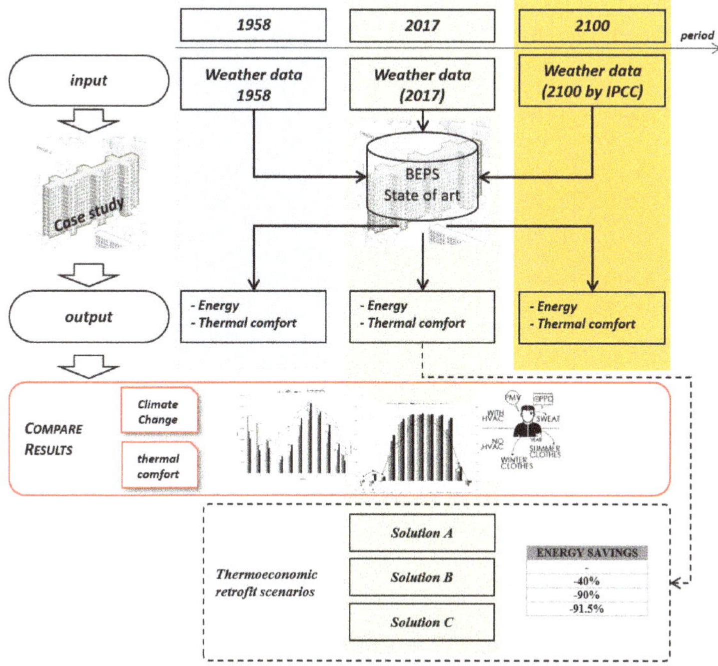

Figure 1. Flowchart of the research methodology.

The first phase of the process is to reach a stable configuration of the virtual model in order to assume the weather data as the key variable and then perform preliminary thermo-physical analyses to evaluate the thermal/energy behavior. With weather data being the key variable, it is possible to compare the different behavior and the energy consumption of the building in the three different scenarios investigating the effects of temperature increase and related impact.

A complete set of thermo-physical analysis is provided for each scenario (S1-1958, S2-2017, S3-2100). The set includes also coherent values concerning loads, energy consumption, and indoor comfort, which are useful indicators to evaluate the behavior of the building with reference to the summer conditions. The methodological approach can be easily adapted to any context and case study.

The investigated outputs include: (a) yearly energy need for heating and cooling (kWh/m^2), (b) total electricity/natural gas (MWh) consumption, (c) total energy consumption, (d) indoor comfort evaluated considering the predicted mean vote (PMV) and the predicted percentage dissatisfied (PPD in percentage, %) when the heating, ventilation, and air-conditioning (HVAC) system is operating, (e) indoor comfort evaluated considering the PMV and PPD when HVAC is not operating and consequently focusing on the clothing relevance.

The geometry and configuration of the building were rendered using Integrated Environment Virtual Environment software (IES.VE [42]), as well as the energy simulation to investigate the potential impacts of temperature increase due to climate change.

3.1. Weather Data for Each Scenario

The research methodology requires three sets of weather data [43] corresponding to the three investigated scenarios (Sc.1 1958, Sc.2 2017, Sc.3 2100). The number 1958 refers to the closest period respect to the year when the building was erected (1939) and then entered in the decade of ordinary operation, 2017 refers to the present time conditions, 2100 refers to the IPCC projection horizon. Weather files referred to 1958 and 2017 were respectively downloaded from Energy Plus (Energy Plus, 2019) and World Meteorological Organization region and Country. Weather data format is a plain text file *.epw (EnergyPlus Weather) and the National Weather Service Forecast Office (National Weather Service, 2019) was used as a typical meteorological year (TMY) [44]. The third set (future projection) is generated based on the 2017 one, assuming an overheating of 1.5 °C dry-bulb temperature, from 2018 to 2100, reported in the IPCC protocol [9], obtained using Element software. This is the main assumption of the entire article: analyze what happens if the IPCC forecasts on temperature are applied to the building performance simulation. Table 1 allows to compare the temperature increase over the years between the three sets. It can be observed that the 1958's winter temperatures were more rigid than the 2017's ones and this is a main effect of the increased temperatures as a consequence of climate change and the heat island effect because, within 60 years, the NYC's built environment has changed drastically.

Energy demand for heating and cooling was analyzed and compared for the three scenarios (climate models) [45,46]. The proposed methodology can be applied to different places simply by replacing the initial climate conditions (.epw) of the site under investigation.

Table 1. Average outdoor temperature.

Year	1958	2017	2100
	Average Outdoor Temperature °C	Average Outdoor Temperature °C	Average Outdoor Temperature °C
January	0.06	3.39	4.89
February	0.12	5.44	6.94
March	3.97	4.18	5.68
April	11.10	13.96	15.46
May	15.27	16.36	17.86
June	19.85	22.20	23.70
July	23.03	24.88	26.38
August	23.55	23.36	24.86
September	19.11	21.45	22.95
October	14.06	17.87	19.37
November	8.55	8.13	9.63
December	2.81	1.70	3.20
Average year	11.79	13.57	15.07
		+1.8 °C compared to 1958	+1.5 °C compared to 2017 +3.3 °C compared to 1958

3.2. Limitation of the Methodology

Research relating to climate change is focused on the effects of energy use, pollution, and environmental issues. The main objective of this paper is not to report the review of these studies, but to highlight the effects that it will have as a result of climate change both on the energy consumption of buildings (specifically for this case study) and for indoor comfort and, consequently, on the clothing insulation. Given the complexity of the research carried out, some methodological limits have been assumed to be able to extend and compare the results with other (future) research. The methodological limits concern:

- The choice to consider the climate change effects only as an increase in the outdoor temperature of 1.5 °C;

- The choice to consider as a parameter for the evaluation of the indoor comfort the internal temperature, thus excluding the detailed analysis of other parameters such as the mean radiant temperature (MRT) or relative humidity (RH);
- The choice to consider clothing insulation as a variable that will suffer the effects of climate change; in this case, the variable adopted is the thermal resistance of the clothing (I_{cl}) (expressed in clo, m^{2}K/W).

The unavailability of original consumption data did not allow the validation of the virtual model according to real conditions, however, a standardized calculation method has been adopted and the same approaches can be adopted in any other case study in a different context.

3.3. BEPS of the State of the Art

Energy simulations were performed using the Integrated Environment Virtual Environment software (IES.VE [42]).The building was modeled to fix the main geometrical features and shape, the boundary conditions, the heating and cooling systems, as well as the services and installations. The building was approximated to a volume 100 meters long, 10 meters wide, and 40 meters tall, corresponding to an affordable housing building hosting 300 apartments in its 14 stories. The study estimated the energy consumption in each scenario with the purpose to evaluate the impact of climate change on the entire building energy demand. The heating/cooling set-points are defined in 20 °C/27 °C, target temperatures within the range that is acceptable to 80 percent of the building occupants according to ANSI/ASHRAE Standard 55, Thermal Environmental Conditions for Human Occupancy. A virtual simulation model was created, and the outcomes of simulations were then compared. Simulations considered the thermal inputs given by internal gains. The same building model (geometry, thermophysics, internal gains, etc.) was used for the simulation purpose, but each simulation was associated to its own weather data file (1958, 2017, 2100) of the set climatic zone: New York Central Park (*.fwt or *.epw file). The increase of temperature due to climate change influences the energy consumption of the building, leading to worsening of comfort conditions.

3.4. Thermal Comfort and Clothes Insulation

The reference model about thermal comfort is still the one developed by Fanger [47] that focuses on two indices: predicted mean vote (*PMV*) and percentage people dissatisfied (*PPD*), which respectively refer to occupants' mean thermal sensation vote and the percent of people voting. The *PMV-PPD* model is a widely used design tool incorporated in thermal comfort standards [48,49] that can be equally applied to different building typologies and climate conditions. Nevertheless, the accuracy of the *PMV-PPD* model in predicting thermal comfort has been questioned through field studies in real buildings [50] as well as in laboratory studies [51].

With this reference, the reported study focused on the relevance of occupants' clothing with relation to their thermal sensations inside the building. Statistical analyses were performed with the aim to better understand how building occupants can achieve thermal comfort by adjusting their clothing level of insulation [52]. This is quite relevant, assuming the poor insulation characteristics of the building envelope and the low-income condition of most of the residents which limit the opportunity to reach adequate comfort levels.

The proposed neutral clothing model can be used to determine whether clothing adjustment can sufficiently offset indoor temperatures in naturally ventilated building contexts [53]. The driving question was therefore: which clothing allows to perceive a good comfort condition comparable to a *PMV* = 0 (neutral) in the scenario without HVAC? To provide an answer for each scenario (1958, 2017, 2010) *PMV* was imposed = 0 and the calculation was repeated by setting different clothing insulation values I_{cl} (clo).

The HVAC system was turned off in order to create a neutral condition, otherwise the system operates to achieve satisfactory comfort conditions.

Clothing insulation (I_{cl}) is an input variable for thermal comfort calculations, and appropriate clothing adjustment can widen the comfort range and reduce building energy consumption [54]. For people in sedentary activities (a metabolic rate of approximately 1.2 met), the effect of changing clothing insulation on the optimum operative temperature is approximately 6 °C per clo, which is much relevant.

3.5. Thermo-Economic Retrofitting Solutions for Improving the Building Energy Performance

The thermo-economic analysis combines thermo-physical analysis with economic aspects to identify factors involved in the generation of energy costs. Highest inefficiencies and costs are related to the processes in which natural gas is burned, hence in the boilers. The presented analysis can be extended to other energy systems. The trend for the following years shows a continuous increase in world population, CO_2 emissions, and primary energy consumption. For that reason, in the building environment, the design of efficient systems has become a primary objective for energy policies in most countries [55]. The starting assumption of the study is that the investigated solutions are expected to let the building reaches a lower consumption, according to the major rating systems, meaning a consumption not exceeding 15–20 kWh/m^2 on an annual basis and the fulfilment of a number of sustainable indicators. Such conditions are achieved through a combination of different design choices involving several parameters, which are not all under investigation, in a balanced contribution deriving from the building envelope characteristics and services/installations.

For the case study under investigation, three main retrofit solutions were considered.

Retrofit A solution envisages the replacement of windows, respecting the U-values thresholds, and adding an insulation layer in the inner side of the wall to maintain the original exposed brick facing of the building.

Retrofit B solution replaces the original natural gas installation for heating and the single units for cooling with an electric heat pump and adding a controlled mechanic ventilation in each blind space.

Retrofit C solution adds to the measures included in A and B solutions, the renovation of the apartments following the guidelines of New York City HPD (Housing Preservation and Development) Department.

The set includes also coherent values concerning energy consumption for cooling and heating, which are useful indicators to evaluate the change of building energy performance due to the renovation process. A first observation of the outcomes, gives the chance to considerer how relevant the installations replacement may be in terms of energy performance. The main goal is not simply to compare the energy consumption of each scenario (A,B,C), that would of course reflect the differences due to the renovation intensity, but to address reflections towards the relation between them and the building energy performance as a relevant concern to be properly taken into account during the retrofitting phase for reducing energetic costs of the entire building and following a more sustainable approach.

4. Case Study Description

The case study is part of a social housing development in Brooklyn (NYC) (Figure 2), known as "Red Hook Houses", owned by the New York City Housing Authority (NYCHA), the major social housing developer of North America, who since 1934, provides affordable and qualitatively sufficient housing for the low-income population of New York. Since 1939, when it was erected, the buildings were never heavily renovated and only after 2012 Hurricane Sandy, some extraordinary maintenance activities were done to install new utility systems due to the flooding event. The reference building (Figure 3) is 14 stories tall and was constructed using a reinforced concrete frame system with an exposed brick cladding. The building counts 300 apartments respecting the minimum air-illuminating ratio.

New York City has a humid continental climate with mean annual precipitation of 127 mm (50 inch) [56]. The 2017 mean annual temperature was 13.58 °C, with 24.88 °C in July and 1.70 °C in December, the hottest and coldest months, respectively. The simulation settings were the same in all the three scenarios: each building model had the same internal footprint, window size and glazing properties, the same HVAC system, internal gains, and infiltration rates, as Table 2 summarizes.

Figure 2. New York City—Red Hook Houses in Brooklyn.

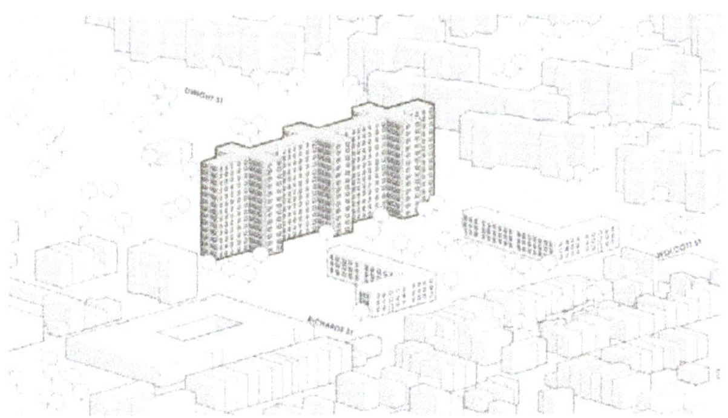

Figure 3. Zoom on investigated building.

Table 2. Building model details.

Building Model Details	Data
Floor Area	1600 m^2 × 14 stories = 22,400 m^2
Orientation	Principal axis running northwest–southeast direction
Windows	One glazed window (1.6 × 1.6 m−1.6 × 1.2 m−1 × 1.6 m−0.5 × 1.6 m)
U-Value	1.6 W/m^{2}K
U-Values (W/m^{2}K)	Walls = 0.16392
	Floor = 0.15853
	Ceiling = 0.10866
HVAC system	Ideal loads
HVAC Set points	20° Heating/27° Cooling
HVAC Schedule	24 h (Continuously on)
Internal Gains	Fluorescent Lighting
	People (max 90 W/person−min 60 W/person)
Infiltration	0.250 ach (max flow)

5. Results

The results of the study are about three different aspects:

1. BEP of different scenarios (1958, 2017, and 2100);
2. Thermal comfort in different scenarios (1958, 2017, and 2100);
3. Thermo-economics evaluations after retrofitting for improving building energy performance.

5.1. BEP Results for Scenarios of Climate Change (1958, 2017, and 2100)

Indoor temperature is one of the parameters used to assess people's health and their comfort level. With the HVAC turned on, the average internal temperature is in a range of 20°C–27 °C all year; when HVAC is tuned off, the temperature varies from minimum peaks of 3°C–4 °C, up to maximums of about 30 °C.

a.   *Heating and cooling energy need loads*

Figure 4 shows the room heating/cooling loads of all scenarios. In the first scenario (1958), the need for heating during the winter months is greater than the other two (2017, 2100), since the average outdoor temperature is lower.

Instead, in summer, the demand for cooling in 1958 was less than the one in 2017 and the one expected in 2100. Compared to 2017, the projection for the summer period always increases while consumption is much higher for heating. Figure 4 shows the relation between energy loads for heating and cooling, by month, compared with outdoor temperature, for each scenario year: 1958, 2017, and 2010. As we can observe, cooling load increase in the 2100 scenario following outdoor temperature, if we compare 1958 and 2100 scenarios, we observe a swift of energy loads from winter (250 MWh heating load in 1958) to summer (320 MWh cooling load in 2100), which is depending on climate change.

b.   *Energy*

Table 3 shows the increase of cooling load despite the decrease of heating loads in all scenarios due to the climate change. Despite this switch, the total amount of consumption remains unchanged. Moreover, Table 3 shows the relation between electricity and fuel use: increasing the temperature, the demand for electricity is higher, while the request for natural gas decreases.

Table 3. Annual average energetic loads (MWh/year).

Energy Data	1958	2017	2100
Heating loads	2220	1825	1475
Cooling loads	1425	1825	2150
Total energy loads	3645	3650	3625
Energy consumption			
electricity	790	835	875
natural gas	185	150	120
Primary energy	975	985	995

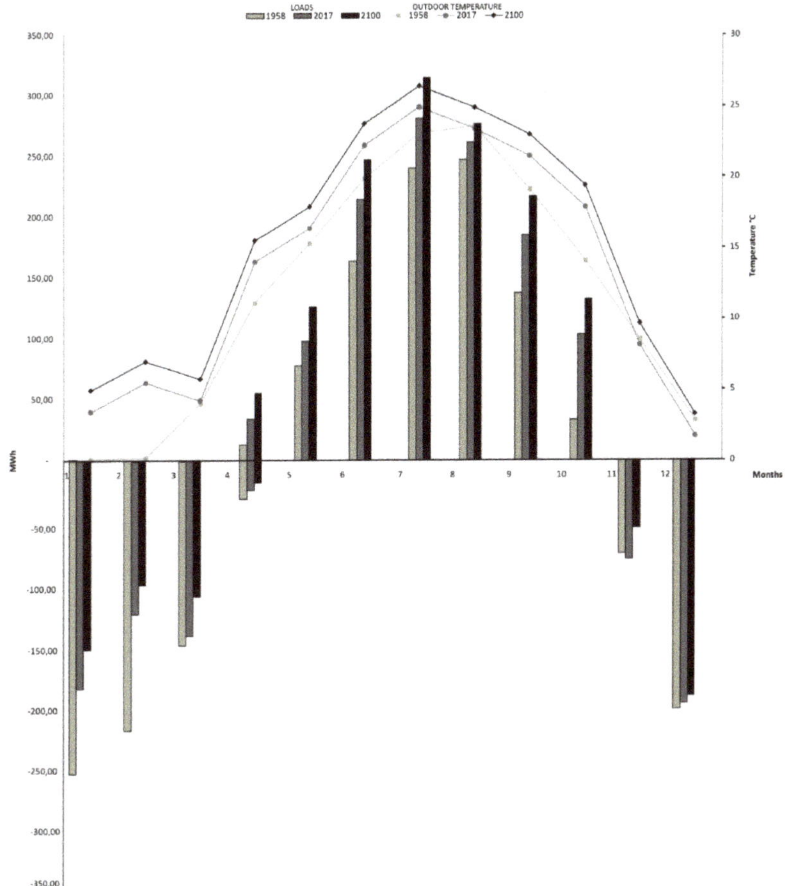

Figure 4. Building energy loads, for heating and cooling, compared to outdoor temperature (°C) in free running.

5.2. Thermal Comfort in the Different Scenarios

The existing clothing models are typically used to estimate a clo-value of a group of building occupants, which then becomes input data for *PMV* calculations following ISO 7730 [48] and ASHRAE 55 [49]. *PMV*-predicted percentage dissatisfied model, thermal neutrality (*PMV* = 0) is deemed as an ideal or optimal status where the number of 'dissatisfied' building occupants is minimum.

a.   *Indoor Comfort WITHOUT HVAC system*

The effect of climate change without HVAC on indoor comfort perceived by users is also investigated. Table 4 shows the *PMV* values (predicted mean vote) of the users when the HVAC system is OFF, correlated to the average external temperature. The results evidence higher value of *PMV*, from −3 (cold) to 3 (hot), meaning that people are dissatisfied. The results prove that installations/services play a very relevant role in users comfort perception; more than 90% of people are dissatisfied during summer.

Table 4. PMV–without HVAC system related to the average outdoor temperature (°C).

Month	1958 °C	PMV 1958	2017 °C	PMV 2017	2100 °C	PMV 2100
January	0.06	−2.35	3.39	−1.82	4.89	−1.52
February	0.12	−2.13	5.44	−1.33	6.94	−1.02
March	3.97	−1.43	4.18	−1.16	5.68	−0.84
April	11.10	−0.08	13.96	0.49	15.46	0.79
May	15.27	1.26	16.36	1.51	17.86	1.81
June	19.85	2.12	22.20	2.49	23.70	2.66
July	23.03	2.79	24.88	3.00	26.38	3.00
August	23.55	2.89	23.36	2.92	24.86	3.00
September	19.11	2.28	21.45	2.57	22.95	2.82
October	14.06	1.01	17.87	1.87	19.37	2.18
November	8.55	−0.28	8.13	−0.34	9.63	−0.03
December	2.81	−1.59	1.70	−1.64	3.20	−1.33
Annual average	**11.79**	**+0.37**	**13.58**	**+0.71**	**15.08**	**+0.96**
Winter average (Oct–Mar)	**4.93**	**−1.13**	**6.79**	**−0.74**	**8.29**	**−0.43**
Summer average (Apr–Sep)	**18.65**	**+1.88**	**20.37**	**+2.16**	**21.87**	**+2.35**

We can observe in Table 4 the annual average value of thermal comfort index *PMV* increase from +0.37 (as "neutral" or "slightly warm" sensation) of 1958 to 0.96 ("slightly warm" nearly "warm" sensation) of 2100, this is a huge difference.

In detail, during the winter season, from October to March, index *PMV* change from −1.13 ("cool" sensation) in 1958, to –0.43 (slightly cool) in 2010, so we suppose a less energy need to heat in the 2100 scenario. During the summer season, from April to September, the index *PMV* change from +1.88 ("warm" nearly hot sensation) to +2.35 corresponds with a very hot sensation, nearly a heat stress value.

These simulations are without HVAC systems, so the sensation depends only by building, so it is evidenced by the main role of heating and cooling equipment and their energy consumption to guarantee a neutral thermal comfort sensation (neutral if *PMV* equal to zero). Following the above approach, not only does HVAC have a role, but also clothes in indoor, described in the next paragraph, we would like to discover which dress people put on their body to feel a neutral thermal comfort, when the *PMV* index is equal to zero.

b.   Relevance of clothing

For ensuring good comfort condition comparable to a *PMV* = 0 (neutral) and metabolic activity is equal to 1 met (seated activity) in no operating HVAC scenario, a *PMV* = 0 was imposed and the calculation was iterated with different I_{cl} (clo) (clothing resistance) for each case. Table 5 provides a summary of the outcomes where it is clearly evident that during the summer period, comfort is acceptable only wearing shorts and a t-shirt (the perception is the one of a clearly overheated environment), while during winter, a polar equipment would be needed to achieve minimum livable conditions.

The adopted approach intentionally stresses the focus on the role of clothing and on the perception of users to remark the dependency of comfort conditions from HVAC and generally from services and installations. This is strictly connected to the inefficient and low performing condition of the building envelope which—at the stating conditions—is highly inadequate to meet the energy demand reduction that is expected in the very near future.

In Table 5, it is possible to observe the annual average value of clothes (or dress thermal resistance) to guarantee a neutral thermal comfort (*PMV* = 0) that corresponds to 2.01 clo (Underwear with short sleeves and legs, shirt, trousers, jacket, heavy quilted outer jacket and overalls, socks, shoes, cap, and gloves, following ISO 9920 Annex A [57]) in 1958 and 1.58 clo (Underwear with short sleeves and legs, shirt, trousers, vest, jacket, coat, socks, and shoes, following ISO 9920 Annex A) in 2100, with a decrease of 0.40 clo of clothes thermal resistance, which corresponds to a sweater or jacket. In detail, during the winter season, from October to March, clothes thermal resistance (I_{cl}) change from 3.22 to

2.67 clo with a gap of 0.55 clo, while in the summer season, from April to September, the gap is 0.30 clo, which corresponds to a t-shirt.

Table 5. Clothes insulation (measured in clo) without HVAC and *PMV* = 0, related to average indoor temperature.

Without HVAC	1958 °C	I_{cl} 1958	2017 °C	I_{cl} 2017	2100 °C	I_{cl} 2100
January	3.61	4.3	6.76	3.5	7.89	3.3
February	4.12	4.0	8.96	3.1	9.94	3.0
March	7.98	3.3	8.45	3.1	8.68	3.1
April	15.03	2.1	17.82	1.6	18.46	1.5
May	20.3	1.2	21.03	1.1	20.86	1.1
June	24.3	0.6	26.57	0.3	26.7	0.2
July	27.75	0.2	29.57	0	29.38	0
August	28.2	0.1	28.04	0.1	27.86	0.05
September	23.92	0.6	26.04	0.3	25.95	0.13
October	18.26	1.6	22.22	1	22.37	1
November	12.32	2.6	11.87	2.7	12.63	2.6
December	6.51	3.5	5.86	3.9	6.2	3.
Annual average	**16.03**	**2.01**	**17.77**	**1.73**	**18.08**	**1.58**
Winter average (Oct-Mar)	**8.80**	**3.22**	**10.69**	**2.88**	**11.29**	**2.67**
Summer average (Apr-Sep)	**23.25**	**0.80**	**24.85**	**0.57**	**24.87**	**0.50**

On the other hand, in Table 6, there is a comparison between the *PMV* index that the users would have if the indoor temperature would have been 20°C with the HVAC system OFF. The results of these simulations are the values of clothes thermal resistance (I_{cl}), which shows how a user might be dressed for having this *PMV* index in this environment.

Table 6. Clothes insulation in a T = 20 °C ambient with a *PMV* index of the HVAC system OFF simulations.

Clothes Insulation without HVAC System	PMV 1958	I_{cl} 1958	PMV 2017	I_{cl} 2017	PMV IPCC	I_{cl} 2100
January	−2.35	0.10	−1.82	0.50	−1.52	0.70
February	−2.13	0.30	−1.33	0.75	−1.02	0.90
March	−1.43	0.70	−1.16	0.90	−0.84	1.00
April	−0.08	1.35	0.49	1.55	0.79	1.65
May	1.26	1.80	1.51	1.90	1.81	2.00
June	2.12	2.10	2.49	2.20	2.66	2.25
July	2.79	2.30	3.00	2.40	3.00	2.40
August	2.89	2.35	2.92	2.40	3.00	2.40
September	2.28	2.15	2.57	2.25	2.82	2.35
October	1.01	1.75	1.87	2.00	2.18	2.10
November	−0.28	1.15	−0.34	1.15	−0.03	1.35
December	−1.59	0.65	−1.64	0.60	−1.33	0.70
AVERAGE	**0.37**	**1.39**	**0.71**	**1.55**	**0.96**	**1.65**

5.3. Thermo-Economics Evaluations Concerning Retrofit Process

Simulation is credited with increasing efficiency and enabling the comparison of a broader range of design options, leading to a more balanced and optimized design. Simulations provide a better understanding of the consequences of design decisions, which are supposed to increase the overall building energy performance. The starting assumption of the study is that the building has to be renovated to meet a 15–20 kWh/m^2 yearly energy demand threshold. Therefore, the renovation work costs, and the investment needed to support the process were analyzed and compared among the different solutions.

Table 7 shows the results of the simulations for each scenario. The current energy performance of the building is 155 kWh/m^2 per year, which is a quite high level of energy demand considering the target.

Table 7. Results of different scenarios.

Solution 2017	Energy Consumption	Energy Savings	Energy Cost	Renovation Investment
ACTUAL	155 kWh/m^2	-	88 \$/m^2	-
Solution A	94 kWh/m^2	−40%	86 \$/m^2	1600 \$/m^2
Solution B	15 kWh/m^2	−90%	27 \$/m^2	1800 \$/m^2
Solution C	13 kWh/m^2	−91.5%	26 \$/m^2	2200 \$/m^2

Solution A—includes the replacement of windows and the introduction of an additional insulation layer in the inner side of the wall—lead the energy performance to 94kWh/m^2 per year, requiring a quite relevant investment for renovation.

Solution B—improves solution A and operates replacing the existing heating/cooling system with an electric heat pump and adding a controlled mechanic ventilation—reduces the energy demand to 15 kWh/m^2 with about 90% of energy saving compared to the current status.

Solution C—improves solution B and simply adds the apartments refurbishment according to the HPD guidelines, achieving an overall energy performance of 13 kWh/m^2 per year.

6. Discussion

6.1. BEP of Current Conditions in Different Scenarios (1958, 2017 and 2100)

Looking at the outcomes of the performed analyses of the three scenarios (1958, 2017, 2100) is evident how loads, consumptions, and comfort are influenced by climate data. The research allowed to understand the variations correlated to outdoor/indoor temperature dealing with the building behavior (loads), the energy data (consumption), and the comfort data (PMV/PPD).

Comparing the situation between 1958 and 2100, reported in Figure 4, it can be noticed that there is an increase in the building's cooling needs of 50% in the 2100 prediction with respect to 2958 (+14% in 2017 with respect to 1958), which clearly addresses the design challenge toward the summer period to find adequate technical answers for reducing the overheating effect. In 1958, energy demand for heating was 19%, while in 2100, it is expected to decrease to 12% (see Figure 4) with a consequent increase in the share for cooling, with a radical shift in the energy use and building thermal behavior compared with a conventional situation. Renovation works will reduce the energy demand for heating (fuel), replacing the existing boilers), but the request for electrical power will increase exponentially, impacting on the energy infrastructure. The increase of outdoor temperature will impact on indoor conditions and consequently on comfort condition for inhabitants and their related perception, which will be dependent on HVAC and installations.

6.2. Thermal Comfort

To understand how thermal comfort varies if HVAC is not operating, the users' perception was investigated with the purpose to better address the solutions dealing with the building envelope main features and performance. The increase in the share of energy demand for cooling is of particular interest at the European level where the use of the HVAC during the summer period is a quite recent need. Historic buildings are usually massive structure with natural ventilation and the wide building production between the 1960s and the 1980s was not originally air-conditioned. This represents a relevant cultural difference with the United States where the use of the HVAC is consolidated over a century. Figure 5 report a infographics about thermal comfort in each scenarios.

The choice to work on the NY case study reflects the will to work on a consolidated situation for both winter and summer energy demand regimes, assuming that the huge increase in the demand for cooling will represent a revolving trend in EU for the future.

The scientific literature provides a large production of studies about building thermal behavior and energy performance as well as about the retrofitting and renovation solution, however, the impacts of the projections in temperature trends reported by IPCC still represent an open field of research.

The novelty of the proposed study lies in considering the interrelated factors focusing on the effects on thermal comfort perception by the users and on the potential deriving changes in life conditions and styles.

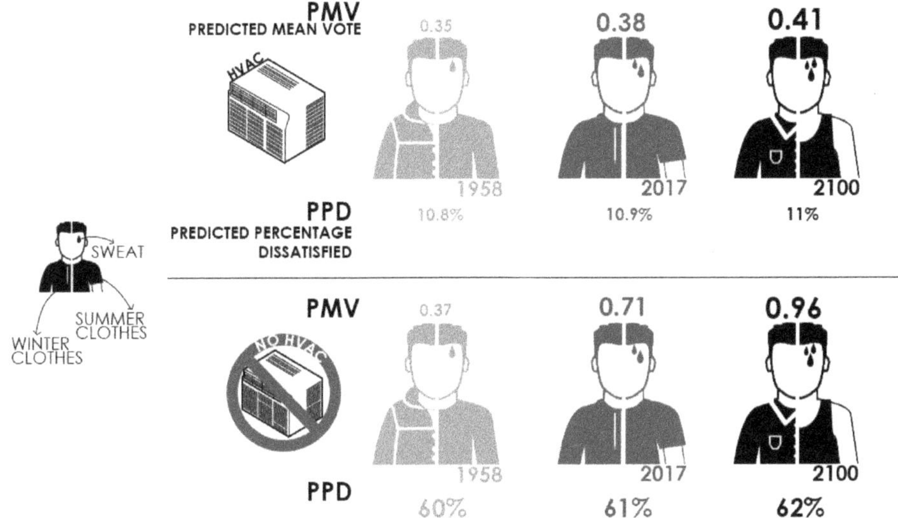

Figure 5. Infographic about thermal comfort in each scenario considering the HVAC system ON and OFF.

6.3. Thermo-Economics Evaluations after Retrofitting for Improving Building Energy Performance

Among the objectives of this study, there is the understanding of the impact of building renovation on heating/cooling demand with relation to the temperature increase projection. Three building renovation solutions were developed with relation to the 2017 scenario. Assuming the current status as the starting point, solution A replaces the existing windows with new ones, respecting the directives U-value (W/m^2K) thresholds and adds a new insulation layer, achieving a savings of 40%.

Solution B replaces the existing boiler for heating with an electrical heat pump for both heating and cooling, combined with a controlled mechanic ventilation, achieving a savings of 90%. Solution C adds to B the refurbishment of the apartments according HPD's guidelines.

The total energy performance shifts from 155 kWh/m^2 (of the current status) to 13 kWh/m^2 per year. The results are obtained modeling only the basic parameters involved such as insulation, window glazing and installations, but other key elements like the level of shading, building orientation, and window size, which were kept constant, can be varied to further improve the outcomes. The study assumed the limitation of preserving the exposed brick façade of the building, but in different context, a more performing insulation on the outer skin can be considered for improving the performances of the building envelope.

7. Conclusions

This research assumes that energy simulation and analysis must be carefully considered and included during the renovation process aimed at improving building energy performances, to achieve effective savings and more efficient design solutions in terms of comfort. The proposed methodology has, as the main objective, the replicability and can be helpful both in the case of renovation and in case of new construction, aiming to consider IPCC projections and the related effects on people's health and wellbeing. The method adopted also allows incremental numerical data to be obtained. This study evidences the risk of grid collapse and blackouts in case of demand peaks for cooling as it happened in NYC in July 1977 and more recently in July 2019, paralyzing the city for more than 3 h. The main answer is, of course, to strengthen the energy infrastructure as the NYC authorities are already doing, however, a modal shift in approaching renovation and comfort evaluation could be of help in supporting a cultural change, which will be an unavoidable consequence of IPCC projections.

The outcomes of the case study confirm that climate change will significantly affect the future improvement of building energy performance and indoor thermal comfort (especially ventilation), becoming a global challenge that will require to work on thermal insulation and reflectance, but also on ventilation parameters that have a very relevant role on the demand for cooling.

That climate change and improvement of thermal and ventilation parameters significantly affect the demand for cooling—in time, it will be a huge global problem.

Further developments of the proposed study are certainly needed to explore the effects on different building typologies as well as on the inhabitant's reaction considering different climate contexts and cultural background.

Author Contributions: Conceptualization, K.F. and J.G.; methodology, K.F and J.G.; software, L.F.; validation, K.F., L.F. and J.G.; formal analysis, L.F.; investigation, L.F.; resources, J.G.; data curation, K.F.; writing—original draft preparation, L.F.; writing—review and editing, K.F. and J.G.; visualization, L.F.; supervision, K.F.; project administration, J.G.; funding acquisition, J.G. All authors have read and agreed to the published version of the manuscript.

Funding: This research received no external funding

Acknowledgments: In this section you can acknowledge any support given which is not covered by the author contribution or funding sections. This may include administrative and technical support, or donations in kind (e.g., materials used for experiments).

Conflicts of Interest: The authors declare no conflict of interest.

Nomenclature

AR5	IPCC Fifth Assessment Report
BEP	Building Energy Performance
BEPS	Building Energy Performance Simulations
BPS	Building Performance Simulations
BLC	Building Life-Cycle
CC	Climate Change
EPBD	Energy Performance of Buildings Directive
EU	European Union
GHG	Greenhouse Gas
HPD	New York City Department of Housing Preservation and Development
HVAC	Heating, Ventilation and Air-Conditioning
IES.VE	Integrated Environment Virtual Environment software
IPCC	Intergovernmental Panel on Climate Change
NPCC	New York City Panel on Climate Change
NYCHA	New York City Housing Authority
nZEB	nearly Zero Energy Building
PMV	Predicted Mean Vote
PPD	Predicted Percentage Dissatisfied
RES	Renewable Energy Sources
RPS	Renewable Portfolio Standard
SB100	California Senate Bill 100
TMY	Typical Meteorological Year
UNFCCC	United Nations Framework Convention on Climate Change

References

1. Tol, R.S.J. Population and trends in the global mean temperature. *Atmosfera* **2017**, *30*, 121–135. [CrossRef]

2. Ciscar, J.C.; Dowling, P. Integrated assessment of climate impacts and adaptation in the energy sector. *Energy Econ.* **2014**, *46*, 531–538. [CrossRef]
3. Schaeffer, R.; Szklo, A.S.; Pereira de Lucena, A.F.; Moreira Cesar Borba, B.S.; Pupo Nogueira, L.P.; Fleming, F.P.; Troccoli, A.; Harrison, M.; Boulahya, M.S. Energy sector vulnerability to climate change: A review. *Energy* **2012**, *38*, 1–12. [CrossRef]
4. Hilden, M.; Huuki, H.; Kivisaari, V.; Kopsakangas-Savolainen, M. The importance of transnational impacts of climate change in a power market. *Energy Policy* **2018**, *115*, 418–425. [CrossRef]
5. Pischke, E.C.; Solomon, B.; Wellstead, A.; Acevedo, A.; Eastmond, A.; De Oliveira, F.; Coelho, S.; Lucon, O. From Kyoto to Paris: Measuring renewable energy policy regimes in Argentina, Brazil, Canada, Mexico and the United States. *Energy Res. Soc. Sci.* **2019**, *50*, 82–91. [CrossRef]
6. Radhi, H. Evaluating the potential impact of global warming on the UAE residential buildings—A contribution to reduce the CO2 emissions. *Build. Environ.* **2009**, *44*, 2451–2462. [CrossRef]
7. Pérez-Andreu, V.; Aparicio-Fernández, C.; Martínez-Ibernón, A.; Vivancos, J.L. Impact of climate change on heating and cooling energy demand in a residential building in a Mediterranean climate. *Energy* **2018**, *165*, 63–74. [CrossRef]
8. Yamineva, Y. Lessons from the Intergovernmental Panel on Climate Change on inclusiveness across geographies and stakeholders. *Environ. Sci. Policy* **2017**, *77*, 244–251. [CrossRef]
9. IPCC. Summary for Policymakers. In *Global Warming of 1.5 °C. An IPCC Special Report on the Impacts of Global Warming*; IPCC: Geneva, Switzerland, 2018; ISBN 9789291691517.
10. Livingston, J.E.; Lövbrand, E.; Alkan-Olsson, J. From climates multiple to climate singular: Maintaining policy-relevance in the IPCC synthesis report. *Environ. Sci. Policy* **2018**, *90*, 83–90. [CrossRef]
11. Devès, M.H.; Lang, M.; Bourrelier, P.H.; Valérian, F. Why the IPCC should evolve in response to the UNFCCC bottom-up strategy adopted in Paris? An opinion from the French Association for Disaster Risk Reduction. *Environ. Sci. Policy* **2017**, *78*, 142–148. [CrossRef]
12. Zhai, Z.J.; Helman, J.M. Implications of climate changes to building energy and design. *Sustain. Cities Soc.* **2019**, *44*, 511–519. [CrossRef]
13. Flores-Larsen, S.; Filippín, C.; Barea, G. Impact of climate change on energy use and bioclimatic design of residential buildings in the 21st century in Argentina. *Energy Build.* **2019**, *184*, 216–229. [CrossRef]
14. Wang, H.; Chen, Q. Impact of climate change heating and cooling energy use in buildings in the United States. *Energy Build.* **2020**, *82*, 428–436. [CrossRef]
15. Shen, P. Impacts of climate change on U.S. building energy use by using downscaled hourly future weather data. *Energy Build.* **2017**, *134*, 61–70. [CrossRef]
16. Moazami, A.; Carlucci, S.; Nik, V.M.; Geving, S. Towards climate robust buildings: An innovative method for designing buildings with robust energy performance under climate change. *Energy Build.* **2019**, *202*, 109378. [CrossRef]
17. Augenbroe, G. Trends in building simulation. *Build. Environ.* **2002**, *37*, 891–902. [CrossRef]
18. Coakley, D.; Raftery, P.; Keane, M. A review of methods to match building energy simulation models to measured data. *Renew. Sustain. Energy Rev.* **2014**, *37*, 123–141. [CrossRef]
19. Attia, S.; Hamdy, M.; O'Brien, W.; Carlucci, S. Computational Optimisation for Zero Energy Buildings Design: Interviews Results with 28 International Experts. In Proceedings of the BS2013: 13th Conference of International Building Performance Simulation Association, Chambéry, France, 26–28 August 2013.
20. Gercek, M.; Durmuş Arsan, Z. Energy and environmental performance based decision support process for early design stages of residential buildings under climate change. *Sustain. Cities Soc.* **2019**, *48*, 101580. [CrossRef]
21. Council of the European Union. *Energy and Climate Package—Elements of the Final Compromise Agreed by the European Council*; Council of the European Union: Brussels, Belgium, 2008.
22. European Commission. *A Policy Framwork for Climate and Energy in the Periodo from 2020 to 2030*; European Commission: Brussels, Belgium, 2014.
23. European Commission. *A Clean Planet for all A European Strategic Long-Term Vision for a Prosperous, Modern, Competitive and Climate Neutral Economy*; European Commission: Brussels, Belgium, 2018.
24. Sánchez-garcía, D.; Rubio-bellido, C.; del Río, J.J.M.; Pérez-Fargallo, A. Towards the quantification of energy demand and consumption through the adaptive comfort approach in mixed mode office buildings considering climate change. *Energy Build.* **2019**, *187*, 173–185. [CrossRef]

25. Campaniço, H.; Soares, P.M.M.; Cardoso, R.M.; Hollmuller, P. Impact of climate change on building cooling potential of direct ventilation and evaporative cooling: A high resolution view for the Iberian Peninsula. *Energy Build.* **2019**, *192*, 31–44. [CrossRef]

26. European Parliament. Directive 2010/31/EU of the European Parliament and of the Council of 19 May 2010 on the energy performance of buildings (recast). *Off. J. Eur. Union* **2010**, *L153*, 13–35.

27. European Parliament. Directive 2018/844 amending Directive 2010/31/UE on the energy performance of building and Directive 2012/27/UE on energy efficiency). *Off. J. Eur. Union* **2018**, *2018*, 75–91.

28. European Parliament. DIRECTIVE 2012/27/EU OF THE EUROPEAN PARLIAMENT AND OF THE COUNCIL of 25 October 2012 on energy efficiency, amending Directives 2009/125/EC and 2010/30/EU and repealing Directives 2004/8/EC. *Off. J. Eur. Union* **2012**, *L315*, 1–56.

29. Andrić, I.; Pina, A.; Ferrão, P.; Fournier, J.; Lacarrière, B.; Le Corre, O. The impact of climate change on building heat demand in different climate types. *Energy Build.* **2017**, *149*, 225–234. [CrossRef]

30. Cellura, M.; Guarino, F.; Longo, S.; Tumminia, G. Energy for Sustainable Development Climate change and the building sector: Modelling and energy implications to an of fi ce building in southern Europe Special Report on Emissions Scenarios. *Energy Sustain. Dev.* **2020**, *45*, 46–65. [CrossRef]

31. Andrić, I.; Gomes, N.; Pina, A.; Ferrão, P.; Fournier, J.; Lacarrière, B.; Le Corre, O. Modeling the long-term effect of climate change on building heat demand: Case study on a district level. *Energy Build.* **2016**, *126*, 77–93. [CrossRef]

32. Thomas, K.A.; Warner, B.P. Weaponizing vulnerability to climate change. *Glob. Environ. Chang.* **2019**, *57*, 101928. [CrossRef]

33. Schulte, R.H.; Fletcher, F.C. 100% Clean Energy: The California Conundrum. *Electr. J.* **2019**, *32*, 31–36. [CrossRef]

34. García Sánchez, F.; Solecki, W.D.; Ribalaygua Batalla, C. Climate change adaptation in Europe and the United States: A comparative approach to urban green spaces in Bilbao and New York City. *Land Use Policy* **2018**, *79*, 164–173. [CrossRef]

35. Energy; Planning New York 2015-State-Energy-Plan. 2015. Available online: https://energyplan.ny.gov/Plans/2015.aspx (accessed on 16 June 2020).

36. New York State of Opportunity Reforming the Energy Vision (REV). 2016. Available online: https://www.ny.gov/sites/ny.gov/files/atoms/files/WhitePaperREVMarch2016.pdf (accessed on 16 June 2020).

37. NYS Department of Public Service Reforming the Energy Vision. *Reform. Energy Vis.* **2014**, *85*. Available online: http://documents.dps.ny.gov/public/Common/ViewDoc.aspx?DocRefId=%7B0B599D87-445B-4197-9815-24C27623A6A0%7D (accessed on 16 June 2020).

38. New York Academy of Sciences New York City Panel on Climate Change 2019 Report Executive Summary. *Ann. N. Y. Acad. Sci.* **2019**, *1439*, 11–21. [CrossRef] [PubMed]

39. Shen, P.; Braham, W.; Yi, Y. The feasibility and importance of considering climate change impacts in building retro fi t analysis Intergovernmental Panel on Climate Change. *Appl. Energy* **2019**, *233–234*, 254–270. [CrossRef]

40. Chan, A.L.S. Developing future hourly weather files for studying the impact of climate change on building energy performance in Hong Kong. *Energy Build.* **2011**, *43*, 2860–2868. [CrossRef]

41. Nik, V.M.; Sasic Kalagasidis, A. Impact study of the climate change on the energy performance of the building stock in Stockholm considering four climate uncertainties. *Build. Environ.* **2013**, *60*, 291–304. [CrossRef]

42. IES Virtual Environment Module Tutorial. 2019. Available online: https://www.iesve.com/ (accessed on 16 June 2020).

43. Roshan, G.; Oji, R.; Attia, S. Projecting the impact of climate change on design recommendations for residential buildings in Iran. *Build. Environ.* **2019**, *155*, 283–297. [CrossRef]

44. Farah, S.; Whaley, D.; Saman, W.; Boland, J. Integrating climate change into meteorological weather data for building energy simulation. *Energy Build.* **2019**, *183*, 749–760. [CrossRef]

45. Karimpour, M.; Belusko, M.; Xing, K.; Boland, J.; Bruno, F. Impact of climate change on the design of energy efficient residential building envelopes. *Energy Build.* **2015**, *87*, 142–154. [CrossRef]

46. Wang, L.; Liu, X.; Brown, H. Prediction of the impacts of climate change on energy consumption for a medium-size office building with two climate models. *Energy Build.* **2020**, *157*, 218–226. [CrossRef]

47. Fanger, P.O. *Thermal Comfort. Analysis and Applications in Environmental Engineering;* McGraw-Hill: New York, NY, USA, 1970; ISBN 9780070199156.

48. ISO. *ISO 7730 Ergonomics of the Thermal Environment—Analytical Determination and Interpretation of Thermal Comfort Using Calculation of the PMV and PPD Indices and Local Thermal Comfort Criteria*; ISO: Geneva, Switzerland, 2005; Volume 2005.

49. ASHRAE 55, Thermal Environmental Conditions for Human Occupancy ASHRAE. 2010. Available online: https://www.ashrae.org/technical-resources/bookstore/standard-55-thermal-environmental-conditions-for-human-occupancy (accessed on 16 June 2020).

50. Humphreys, M.A.; Fergus Nicol, J. The validity of ISO-PMV for predicting comfort votes in every-day thermal environments. *Energy Build.* **2002**, *34*, 667–684. [CrossRef]

51. Yang, Y.; Li, B.; Liu, H.; Tan, M.; Yao, R. A study of adaptive thermal comfort in a well-controlled climate chamber. *Appl. Therm. Eng.* **2015**, *76*, 283–291. [CrossRef]

52. Wang, L.; Kim, J.; Xiong, J.; Yin, H. Optimal clothing insulation in naturally ventilated buildings. *Build. Environ.* **2019**, *154*, 200–210. [CrossRef]

53. Liu, W.; Yang, D.; Shen, X.; Yang, P. Indoor clothing insulation and thermal history: A clothing model based on logistic function and running mean outdoor temperature. *Build. Environ.* **2018**, *135*, 142–152. [CrossRef]

54. De Carli, M.; Olesen, B.W.; Zarrella, A.; Zecchin, R. People's clothing behaviour according to external weather and indoor environment. *Build. Environ.* **2007**, *42*, 3965–3973. [CrossRef]

55. Sangi, R.; Martín, P.M.; Müller, D. Thermoeconomic analysis of a building heating system. *Energy* **2016**, *111*, 351–363. [CrossRef]

56. Bader, D.; Tryhorn, L.; Degaetano, A.; Rosenzweig, C. Climate Risks. In *New York State Energy Research and Development, Responding to Climate Change in New York State (ClimAID)*; New York State Energy Research and Development Authority: New York, NY, USA, 2014; pp. 15–48. Available online: https://www.nyserda.ny.gov/-/media/Files/Publications/Research/Environmental/EMEP/climaid/ClimAID-Report.pdf (accessed on 16 June 2020).

57. ISO. *ISO 9920—Ergonomics of the Thermal Environment—Estimation of Thermal Insulation and Water Vapour Resistance of a Clothing Ensemble*; ISO: Geneva, Switzerland, 2008; Volume 2008.

Article

A GIS-Based Methodology for Speedy Energy Efficiency Mapping: A Case Study in Bologna

Jacopo Gaspari *, Michaela De Giglio, Ernesto Antonini and Vincenzo Vodola

Department of Architecture, University of Bologna, Viale Risorgimento 2, 40136 Bologna, Italy; michaela.degiglio@unibo.it (M.D.G.); ernesto.antonini@unibo.it (E.A.); vincenzo.vodola2@unibo.it (V.V.)
* Correspondence: jacopo.gaspari@unibo.it

Received: 12 March 2020; Accepted: 22 April 2020; Published: 3 May 2020

Abstract: The paper reports a methodology developed to map energy consumption of the building stock at the urban scale on a GIS environment. Energy consumption has been investigated, focusing on the shift from the individual building scale to the district one with the purpose of identifying larger homogenous energy use areas for addressing policies and plans to improve the quality and the performance levels at the city scale. The urban planning zoning concept was extended to the energy issue to include the energy behavior of each zone that depends on the performance of its individual buildings. The methodology generates GIS maps providing a district scale visualization of energy consumption according to shared criteria. A case study in Bologna city (Italy) is provided. In the specific case, the last update of Emilia-Romagna regional urban planning regulation required a mapping action regarding energy efficiency of homogeneous urban portions defined by the General Urban Plan. The main achieved results are (a) a methodology to identify homogeneous areas for analyzing energy consumption; (b) an updated energy map of Bologna Municipality.

Keywords: energy efficiency mapping; energy zoning; energy performance certificate; geographic information system; GIS-based methodology

---

1. Introduction

During the last ten years, a number of regulations were introduced at national and local levels across Europe to acknowledge the EU directives concerning energy efficiency [1,2] and to boost energy savings measures in each member state. In addition to the Italian national regulation, Emilia-Romagna Region has recently updated the Regional Urban Planning Law (LR 24/2017) that states each city of the territory is required to perform a mapping action to detect the low quality stock of existing buildings, particularly the ones not fulfilling the minimum thresholds of energy efficiency and seismic safety (art. 22, par. 6). The request is not associated with any guidelines for performing the mapping action, leaving each municipal administration (MA) free to adopt its own methodology and to display the outcomes accordingly. This makes the possible results often very heterogeneous and hard to compare, depending not only on the adopted approach but also on the quality and reliability of the data to be processed.

This paper summarizes the outcomes of a study that was commissioned by the City of Bologna to define a comprehensive and possibly replicable approach to answer the request while achieving a useful methodology to address the planning actions of the near future, taking into account the energy issue as a key priority in the city development. The proposed approach aims to provide a solution for a quick energy mapping with the purpose of reducing the workload usually needed to face a very time-consuming activity that deals both with dedicated data collection and processing. The method therefore considers the use of already available datasets. As evidenced in the literature [3–8], data processing could take an extremely long time, due to the different data collection criteria and above all to the frequent lack of long-term plans for data collection campaigns within the public administrations. This has led to reflect on how mapping is usually conducted, according to the purpose. In most cases, mapping actions are generally based on a precise collection of data related to very appropriate basic units in order to increase the reliability level. However, in this specific case, this would drive the approach to read phenomena at the building scale, rather than at the city scale (which is the typical level of planning actions). Consequently, this study approaches the mapping action accepting some limitations and approximations with the objectives of accelerating the process and providing a comprehensive picture (which can be appropriately updated) in a timely manner of the energy demand at a broader scale.

In order to meet the Bologna Municipality requests, this research looks at energy efficiency from a different perspective, shifting the scale of investigation from the building to the district scale. Energy efficiency is traditionally surveyed at the building scale by analyzing the building envelope, the technical installations and the whole-building energy performance. However, this does not reflect the scope of the Regional Law request, which is basically oriented to understand how energy efficiency is distributed within the urban fabric and to reveal and the energy consumption trend. Thus, the study associates energy mapping with the zoning concept that is traditionally used in the urban planning disciplines [9,10]. This requires relating energy demand level to each city block, identified as the minimum unit of the mapping action. The scope is to visualize an energy zoning that, reflecting the current trend of energy demand of the city blocks, will allow MAs to improve their basic knowledge level and to increase the effectiveness of the possible actions to be taken in the near future through the urban regulations. According to this scope and general framework, the research initially investigated input data availability. This was particularly challenging due to the different data sources and related quality as well as to the different update levels of the datasets produced over time.

The outcomes of the study are (a) the definition of a methodological approach for energy mapping based on the energy zoning concept and (b) the visualization of energy zoning of the Bologna city case study via GIS environment.

The study represents a pilot experience that, despite facing several limitations and barriers, tries to offer an original approach to the topic, focusing on a broader vision to support future planning actions and defining a methodological backbone that can be refined and integrated with new data to improve its quality and reliability.

2. Background Knowledge and State of the Art

Despite the scientific literature offering a consolidated knowledge on surveying of energy efficiency at the building scale, experiences at district or city scales are still limited and are mainly associated—at least across EU—to wide research projects funded under the umbrella of EU framework programs. Among them, "STEP-UP Strategies Towards Energy Performance and Urban Planning" is one of the first outlining the need to focus on large-scale visions connecting initiatives to support renewable energy source (RES) use with demand trends and player engagement [11]. Dedicated studies operating at the city scale are provided by Caputo et al. [12], Theodoridou et al. [13] and Heiple et al. [14], among others [15].

However, any vision at the urban scale requires data regarding energy consumption to be collected, processed and then displayed with relation to a city map that is typically a georeferenced queryable graphic representation in a geographic information system (GIS) environment (as happens in many other sectors, such as mobility and green infrastructure). According to the most shared definitions, georeferencing is the process of assigning real-world coordinates (i.e., related to an official geographic or cartographic reference system) to a specific object that can be a pixel of raster data [16] or a vector entity such as a point, a line or a polygon [17–19]. All possible input data have to include coordinates or linkable information to available georeferenced geometries, such as cadastral data or street name and building number [20]. The objects can differ in size or typology, affecting the level of detail of the investigation (buildings, cadastral parcels, blocks, etc.). The level of detail consequently has to be set according to the mapping purpose and based on the available input data, which must be reliable and accessible. An overall data analysis is needed to set the process and ensure that the outcomes can be clearly understood by the users [21], while adhering to the current regulations regarding privacy in terms of data protection (as data can potentially give information on users' behaviors or living preferences) [22].

GIS is often used in the energy domain to visualize consumption according to specific parcels or filtering data by building typologies [23,24], to address retrofitting actions [21,25], to predict the potential economic and environmental impacts of the adoption of RES use in a specific context [20,26], to plan specific energy infrastructure initiatives [27] or to obtain a projection of atmospheric emissions due to energy consumption [28]. Some studies use GIS technologies to provide energy mapping of built units (individual buildings, blocks, urban fabric portions, etc.) [25,29], offering good evidence of its effectiveness for the purpose and ensuring a precise data georeferencing, a prompt update over time and an easy way to visualize complex information. Since the entailed energy flows are the basic input data needed to draw a GIS map of an area's energy efficiency [23,30,31], the availability and the level of detail of data regarding the energy performance of each considered entity affects both the precision of the model and its reliability [32–34]. According to Lin [25], two main sources can be theoretically used for the scope: on the one side, real energy consumption data can be gained by the energy supplier metering system, assuming that real-time metering devices are installed and that there is a cooperation agreement with the supplier; on the other one, energy consumption can be extracted from energy bills, assuming that users are able to provide this information [32,35]. Unfortunately, there are at least two main critical barriers that obstruct the concrete application of this theoretically optimal approach.

(1) Specific or individual information and data about energy consumption are confidential, and any potential use is regulated by personal data protection [36]. Consequently, they are not easily accessible without the massive and complex involvement of all users potentially involved in the area under investigation [22].

(2) Aggregated data can be more often accessible (under request to the responsible entity); however, the way in which data are aggregated into sets may differ from the physical aggregation of the involved building unit or may be recorded at a scale which does not fit the purpose of the study [37,38].

According to Tronchin and Fabbri [39], energy consumption can be alternatively obtained from energy performance certificates (EPCs), which must be associated to any new or renovated building unit, reflecting the predicted average energy demand based on the technical and constructive characteristics of the building unit according to an official calculation code [40]. EPC data may not exactly match real consumption patterns, and eventual deviations have to be verified by sample tests. Despite being an indirect way to estimate energy consumption, this is a good and more easily accessible option for gaining input data, even if it might not cover the whole stock included within a territorial administrative area. Accurate and systematic information regarding the energy behavior of the building stock including large shares of old and very old buildings is often unavailable in most European urban areas.

Since comprehensive knowledge about energy demand at the urban scale is a strategic element for implementing urban policies in many sectors, several indirect estimation methods have been developed to overcome the lack of direct consumption data. Among them, TABULA—which is the result of an EU-funded project under the IEE program 2009–2012 [41]—established a harmonized catalog of building typologies for the European residential building stock, assigning a typical energy consumption level to each identified category. The typological classification criteria are based on the age of the building, the number of floors and some constructive features that can be identified quite easily by a speedy survey, allowing the application of the methodology to a wide number of cases and keeping the reliability of results within acceptable levels of approximation [42,43]. A good coverage of the required data concerning the involved building stock is an essential factor in limiting the unavoidable imprecision rate of this indirect methodology. Since the results typically refer to the energy demand of a single building, homogeneous coverage is even more relevant when the main objective is to analyze the energy consumption at the scale of aggregated buildings, where potential gaps may seriously affect the reliability of outcomes [44].

3. Data Collection and Methodology

Even though the present research was generated by a specific request at the regional level, the study was developed within a general theoretical approach with the purpose to offer a "speedy method" to allow MAs to consider the energy performance of the urban stock at a broader scale that can possibly be replicated in different contexts. As evidenced by the scientific literature consulted, a shift from the building to the urban scale in energy consumption analysis is required to better address the future actions in city development and transition to a low-carbon environment.

The study is based on the idea of translating the energy zoning concept into energy maps—to be typically displayed at the district scale—with the purpose of obtaining a tool for easily visualizing energy consumption trends and distribution within the territory under investigation. Accordingly, the GIS environment is considered a powerful way to correlate data and information under investigation with georeferenced data. The GIS environment allows the data to be easily updated and enables the them analysis outcomes on be displayed on the city map.

The methodology was based on two main types of input data strictly related to the building entity that—being the object to which energy performance/demand is associated with—was assumed as the basic unit of the process:

(a) The first required input data typology deals with vector files of the geometrical units to which energy data correspond to. This is a compulsory element needed to locate the basic building unit in the map and univocally associate the related energy data within a GIS environment.

(b) The second required input data typology regards energy consumption of the basic considered units. It has to be noted that this unit may differ from the building as a whole, according to the typology and function.

3.1. Data Collection

Vector files have to possibly cover the whole city administrative territory (or at least the one within the investigation boundary). The vector files shall refer to the geometric definition of the buildings, the cadastral parcels, the city blocks, etc., including a set of related topographic information. Generally, the basic unit is associated to individual buildings. These vector files are typically owned by the City Technical Offices for many different purposes and created in different phases according to the need and the scope. It is relevant to note that they frequently present some discrepancies depending on the way in which geometrical data were collected and the files generated. However, considering the whole scale of the process, this can be considered a minor issue.

Energy consumption data represent the key variable of the process and, methodologically speaking, real data are certainly to be preferred, where available, in order to increase the reliability of outcomes. As Figure 1 shows, the availability or unavailability of real data determines the steps of the methodology. Unfortunately, real data are not so frequently available and accessible, as the ownership in most cases is in the hands of the energy supplier, who has to comply with data protection regulations.

Even though specific agreements can solve these limitations (as has happened in some cases), past experiences and the literature [45–48] suggest considering real data to be unavailable. The present study aims to face the most recurring situation where many barriers obstruct the use of these data; thus, keeping in mind the speedy approach of the method, a worst-case scenario is considered. An alternative source of input data is therefore required. Energy performance certificates (EPCs) represent a viable solution. When approaching this data source, the distribution within the city administrative territory has to be carefully checked to make the process feasible. Once input data are secured, the linkage through coordinates or georeferenced information in the GIS environment must be explored, and the coverage within the territory under investigation must be quantified.

Assuming the use of EPCs as primary data source related to energy demand, the linkage to the vector polygon must be consistent with the administrative entity, usually the cadastral parcel, with EPCs being administratively associated to the building property (which typically refers to the cadastral system). It has to be noted that frequently the cadastral parcel does not perfectly match the building geometrical definition, and the vector files may consequently differ. In the case of more units being included in one building or cadastral parcel, EPCs can be mediated considering the coverage and the boundaries of the selected unit. The decision regarding the file to be used depends on quality of the available files made available by the MA.

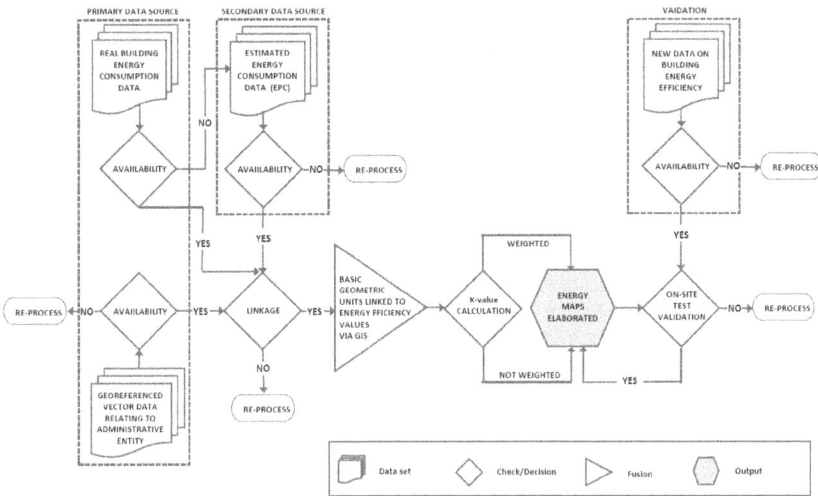

Figure 1. Methodology workflow.

3.2. Methodology

The methodology requires obtaining a unique shapefile embedding all original or mediated energy efficiency values univocally associated to the selected basic unit (buildings, cadastral parcels, blocks, etc.).

As the main objective is not to provide a picture of a constellation of individual building energy performance but rather to focus on detecting homogeneous or heterogeneous areas in energy demand within the city territory, a shift from the building or the cadastral parcel geometry to the one of the city block is required. An intermediate required step is to properly define the city block. The definition comes from the urban planning discipline trying to overcome some semantic differences attributed to the urban basic units country by country. Thus, the city block is intended as an ensemble of buildings and in-between spaces (courtyards, private passages, gardens, etc.) delimited in the perimeter by the typical public viability and representing the typical urban unit of larger neighborhoods or districts.

The shift from the basic unit data association to the city block is operated calculating the so-called K index, which represents an average value obtained from the summation of the energy performance of all the available EPCs included in the city block (mediated according to the related usable area) over the city block surface. The K index is initially calculated as the weighted average of the total primary energy (Ep_{tot}) needs of the building units belonging to the same cadastral parcel, with a weight equal to the useful surface. In the case of generalizing K-values from building level to block level, significant differences may occur in the building size and typology within the same block, and the area/size of a building can be used as a weight for calculating the average K-value.

Subsequently, the K-value at the block level is obtained from the simple average of the K index, previously calculated at the building level (or the cadastral parcel level). Therefore, K is intended to be a geometrical correlation factor between the cadastral parcel unit (or the selected basic unit) and the EPC database (which is also based on a geometrical definition). This allows easily updating the aggregated index if additional data become available within each city block reflecting the improvements made at the individual building scale. At the same time, this allows associating an average behavior with a larger portion of city, providing the MA with a broader picture of the situation sector by sector.

The outcome is a vector map showing the energy efficiency associated to each polygon. The vector data include an attribute table containing all the information related to the considered geometries, which allows extracting specific information related to particular conditions. The map may include some unclassified polygons, due to the eventual lack of input data regarding the EPCs (cadastral data, geometric data), that appear as neutral but may be updated when the related input data become available. The energy use intensity can be associated to other maps (typologies, functions, density, etc.) in order to better define specific improvement and development policies, depending on the MA programs.

The novelty of the proposed methodology does not deal with data processing—as the intention was to adopt currently available tools to facilitate the replication and adoption of the process—but lies in the approach to mapping (from the building level to the block level) considering the urgency of a broader vision and the need of a less time-consuming process to support decision-makers and planners in driving measures at the city level. The contributions offered by the proposed methods to those operating in this sector deal with being a feasible process based on available resources in a quite speedy way, a viable solution to overcome frequent existing barriers (e.g., limited access to real data, limited data coverage, limited precision of vector files) and a replicable process in different contexts using quite comparable input data. The proposed methodology reflects a very flexible approach, designed to be easily replicated, since both the input data processing and the scale of representation can be modified, consequently adapting the process to the local framework in order to achieve comparable maps displaying the available data content [49].

Even though the methodology may suffer from some approximations and loss of data (mainly due to lack of a complete coverage), it is designed to easily accept adaptations or updates. The increase of data availability and accuracy can improve the quality of outcomes and mapping process definition, giving the MAs a tool to quickly monitor the mid- to long-term progress and impacts of the adopted policies.

In order to validate the process and the obtained map, different input data must be assumed [20] that possibly adhere to real energy consumption. Considering the abovementioned limitations and the complexity of a large-scale process, the validation step can be performed on sample test sites where information and input data can be more easily accessible and available. A case study application has been performed within the city of Bologna.

4. Bologna Case Study Application

According to the request of the MA, the defined methodology has been applied to the case study of Bologna, a medium-large-size city located in lower northern part of Italy (Figure 2). With approximately 400,000 inhabitants (1 million considering the whole metropolitan area), Bologna is the seventh most populated Italian city [50]. As the capital city of Emilia-Romagna Region, it has an extent of approximately 140 km^2 and a density of 2780 inhabitant/km^2 [50] Bologna is renowned for having one of Italy's largest and most well preserved historical city centers.

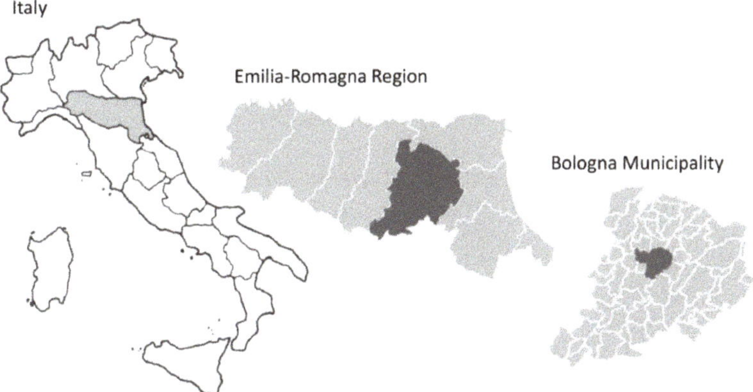

Figure 2. Bologna Municipality (Italy) location.

The urban plan adopted in the late nineteenth century drove the expansion of the city outside the medieval wall ring influencing the city development since the end of the Second World War, when large portions of urban fabric mixing residential and industrial settlements were newly built on the north, east and west sides, often creating dense districts [51–53]. A new general urban plan was adopted in the early 1960s to address further expansions and to mitigate the lack of green and district infrastructure. During the last twenty years, the city of Bologna was characterized by a less intensive building activity more oriented to meet the goals and the key elements of sustainable construction and therefore acknowledging the national regulations that were introduced to adhere to the EU directives. Furthermore, in 2015 Bologna adopted the so-called Bologna Adaptation Plan as the result of the BLUE AP project (Bologna Local Urban Environment Adaptation Plan for a Resilient City), funded by the EU under the umbrella of LIFE+ program (LIFE11 ENV/IT/119), to address future development actions towards more resilient and less energy-consuming solutions to cope with climate change [54].

The update of the Regional Law on Urban Planning (LR 24/2017) therefore represents the last step of a process deeply focused on properly facing the challenge of greatly decreasing energy demand in the coming decades. The study reported in this paper represents the outcome of an intense dialogue with the Bologna administration during the last year and a half.

4.1. Preliminary Phases

During the preliminary phase, the research team and the MA discussed the possible data sources according to availability and accessibility. Real energy consumption data were assumed as unavailable, as this data typology is not owned by the MA and there is no specific agreement with the energy supplier to access it. The initial request of the MA was to estimate the energy performance level considering the year of building construction and both the surface (S) and the volume (V) of each building. The first attempt of association using the TABULA database [55] gave unsatisfactory results, demonstrating that the proposed methodology requires input data regarding energy consumption. Once this was clarified and the MA commitment in this direction was obtained, some key priorities were defined: (a) the level of knowledge about energy performance had to be explicitly related to the basic unit in terms of input data to be made available; (b) a stable and possibly univocal definition of the polygon, corresponding to the building unit, had to be achieved.

Thus, EPCs were assumed as input data regarding energy consumption, being aware that they could not completely cover the whole building stock but having in mind that they are continuously updated and implemented, allowing the methodology to be constantly refined and increased in coverage when new input data become available.

The MA agreed to provide the most updated vector files, regarding both the buildings and the blocks, as well as the EPCs concerning the buildings located within the city's administrative boundaries.

The EPCs files were collected in separate sets, organized according their issuing period, namely 2009–2015 (EPCs from 2009 to 1 October 2015—81,350 records), 2015 (EPCs from 1 October 2015 to 31 December 2015—3265 records), 2016 (full year—11,649 records), 2017 (full year—9973 records) and 2018 (full year—5624 records). The distinction within 2015 is due to the introduction of a new regulation in force from 1 October 2015 (Ministerial Decree 26/06/2015) concerning [56] (a) new elements (such us aeration, air-conditioning, artificial lighting) to be included within the Global Energy Performance index calculation (later called Ep_{tot}) and (b) the use of kWh/m^2 * year as energy performance unit of measurement for all building typologies (while before 1 October 2015, the index was referred to as the volume unit (KWh/m^3 * year) for nonresidential buildings (called E1, including several subclasses)). Due to data protection regulations, EPCs were provided without including the building address and street number, only including cadastral data for GIS linkage purpose.

An EPC database, with cadaster and full location files, was provided to the Municipality by the Emilia-Romagna Region (who is the subject in charge of collecting these data) for validation purpose in sample areas only and with a strict nondisclosure agreement. Thus, the entire first stage of the study was carried out referring to cadastral data only.

4.2. First Stage

The first stage of the process aimed at computing the energy efficiency K indexes to be associated with the map geometries. This index refers to the Ep_{tot} values and the total usable area belonging to each individual cadastral parcel. The index is thought to mediate the potential lack of some EPCs in the same cadastral parcel (the lack could be amended in the future when the missing data become available). The calculation process allows monitoring the loss of data in each cadastral parcel. The K indexes were calculated based on EPC data and using Microsoft Excel and QuantumGIS, according to the following steps:

(1) Each EPC dataset was filtered to delete records (almost 3%) including errors (i.e., zero or missing value for heated volume field, usable area field, cadastral parcel field, or formally incorrect values, etc.)

(2) The EPC datasets were merged into a single file (111,861 records), detecting and erasing EPC duplicates (almost 1%) associated to the same building unit to keep the most updated one.

(3) Ep_{tot} values determined before 1 October 2015 for E1 building class were converted to express the unit measure in usable floor area instead of heated volume (this is to amend the change in the regulation that occurred).

(4) Ep_{tot} values (almost 1.5%) over the thresholds of 500 KWh/m^2 * year for residential buildings and 800 KWh/m^2 * year for nonresidential buildings were excluded, assumed to be out of scale based on regional guidelines and the literature [57]. At this stage, the total number of EPCs suitable for K index calculation shrunk to 106,469.

(5) As reported in the literature [58–63], most of energy efficiency indexes take into account the building envelope characteristics, the heating/cooling and services systems and the concurrent features in order to determine the energy demand. EPCs comply with calculation standards, and the proposed K index aims to mediate the sum of energy performance of the basic units within each cadastral parcel. The energy efficiency K index was calculated as follows (Equation (1)):

$$K = \frac{\sum\left(Ep_{tot} \times \text{Usable area}\right)}{\sum \text{usable area}} \tag{1}$$

where K (KWh/(m^2·year)) is the energy efficiency index, Ep_{tot} (KWh/(m^2·year)) is the energy efficiency value for single unit with EPC, $\sum$ usable area is the total usable area (m^2) or the sum of all the useful surfaces with available EPCs within the same cadastral parcel and Usable area (m^2) is the single unit floor surface.

(6) K index values were checked to delete eventual errors (less than 1%).

The obtained K indexes computed for each cadastral parcel were then ready to be transferred to the GIS environment. For this purpose, the merged EPCs file (steps 2–4) with 106,469 records, the cadastral parcel vector data (73,984 polygons) and the building vector data (41,010 polygons) were used.

The obtained K index can be consequently associated directly to the cadastral parcel or to all the buildings included in it.

In the first case, a one-to-many relationship was established between the cadastral parcel attribute table (A—father) and the EPC table (B—sons). As the primary key is a unique identifier, the foreign key can reliably reference it as one record [64]. The A primary key was obtained by joining the cadastral sheet and the cadastral parcel fields.

In the second case, a new one-to-many relationship was added. It was established between upgraded cadastral parcels vector (A) and building vector (B). All entities to which an EPC is not associated (such as churches, historic monuments, kiosks, etc.) were previously excluded.

Finally, to add the K index field to building vector, the join function was applied between the building attribute table and the cadastral parcels attribute table.

4.3. Second Stage

In order to adapt the model for urban planning purposes, the energy consumption trends were projected at the urban fabric level in a second stage. The coefficient K was mediated at the block scale. The block is defined as a portion of urban fabric surrounded by public streets whose boundaries are classified from an administrative point of view [65]. However, other scale options might be considered in different contexts according to specific characteristics and goals. The decision was discussed and agreed upon with Bologna MA. This does not limit the replication elsewhere nor the validity, as both are mainly dependent on the selected data source availability. The file containing the geometrical definition of the blocks was provided by Bologna MA. The process was managed by QuantumGIS software according to the following steps:

(a) The block identification code was associated to each building by using the position tool.

(b) For each block, the mean K-value (starting from K-values of buildings included of the same block) was calculated (K_{BLOCK} index).

(c) The new one-to-many relationship was accordingly applied. It was established between block vector (A) and building vector (B), so that a single record of the block attribute table (A) matched many rows in building attribute table. Then, the join function was applied between the block attribute table and the building attribute table to add the K_{BLOCK} index field to the block vector.

(d) Finally, the EPC number used in the calculation of the K_{BLOCK} index for each individual block was calculated. Consequently, the information was transferred to the map using the centroids of the polygons corresponding to the blocks.

As explained in the methodology paragraph, the area/size of a building can be used as a weight for calculating the average K-value, and the option to calculate a weighted average K-value was indeed considered in an early stage. A test in a sample portion leading to a difference of less than 5% was discussed with Bologna MA who explicitly required not to weight the K-value, keeping it directly linked to the EPC values of the units included in the block. Even though the EPC availability may not cover all buildings in the block at the moment, Bologna MA and Emilia-Romagna Region are boosting the update of EPCs; therefore, the integration of new certificates will certainly increase the accuracy in the future. Thus the EPC coverage can quickly reach 100% in the coming years, reflecting the situation within the block and the buildings and allowing the MA to monitor the progress in this direction. After a long discussion with the MA about which option would better fit the specific case, the proposed methodology was confirmed.

4.4. Validation

In order to validate the energy maps generated by the described methodology, a sample area was selected to apply the assessment procedure. The site is named "Bolognina" and is made of large blocks, mostly arranged according to a very regular grid of plots, located in the north side of Bologna, near the railway station. As real data on energy consumption are unavailable (typically owned by energy suppliers and strongly covered by data protection regulation), the validation used a new set of EPCs that were provided for this stage only with the full location address (street name and the street number) for each individual building unit. Similar to the previous stage, data are organized in the following separate files: 2009–2011 (full years), 2012–2013 (full years), 2014–2015 (EPCs until 30 September 2015), 2015–2019 (EPCs from the 1 October 2015). Consequently, the following preprocessing steps were needed:

(1) Each EPC dataset was filtered to delete records (almost 4%) including errors (i.e., incorrectly written values in the Ep_{tot} field and in address or street number field, etc.)

(2) The EPC datasets were merged into a single file (126,383 records), detecting and erasing EPC duplicates (almost 1%) associated to the same building unit to keep the most updated one.

(3) Ep_{tot} values determined before 1 October 2015 for E1 building class were converted to express the unit measure in usable floor area instead of heated volume.

(4) Ep_{tot} values (almost 1.5%) over the thresholds of 500 KWh/m^2 * year for residential buildings and 800 KWh/m^2 * year for nonresidential buildings were excluded, assumed to be out of scale based on regional guidelines.

Then, the validation procedure for the Bolognina sample area was performed, including the following steps:

(1) The specific EPC records were extracted according to Bolognina street locations. The address line was corrected using semi-automatic procedures to make the addresses consistent with those reported in the street number vector file. Address duplications within the street number vector file were deleted to obtain unique values for this field.

(2) The Ep_{tot} mean value was calculated starting from EPCs having the same complete address. This step makes the validation comparable with the procedure used in the calculation of the K indicator.

(3) A new one-to-many relationship was created between street number vector (A) and EPC table (B) to match a single record of the street number attribute table (A) with many rows in EPC table. The join function was then applied between the street number attribute table and the EPC table to add the Ep_{tot} mean field to the street number vectors.

(4) The Ep_{tot} mean value between the street numbers included in the same block was calculated within the upgraded street number vector.

(5) The Ep_{tot} mean block value, representing all the street numbers included in the same block, was associated to the block polygons through the attribution for position tool.

5. Case Study Results

The described methodology and related procedures produced vector files containing all the necessary information to represent the estimated energy consumption values on a GIS-based map. Graphical representation can be displayed according to three different scales: buildings, cadastral parcels and blocks. In order to improve map readability, a color palette that varies from green to red—associated to the best and the worst conditions, respectively—was applied. The color scale was divided into six classes from 0–100 KWh/(m^2·year) to greater than 300 KWh/(m^2·year) (Table 1). The cartographic reference system of all maps is ETRS89/UTM 32N. The original building vector file contained 41,010 buildings and represents the whole city territory at building scale definition.

Table 1. Consumption classes according to the energy map visualization.

	Class	K (kWh/m^2·Year)
1		0–100
2		100–150
3		150–200
4		200–250
5		250–300
6		> 300

The uncolored buildings are those not useful for the scope (such as canopies, abandoned or dismissed buildings, kiosks, barracks, etc.) or belonging to specific categories like historical monuments or those not included due to data unavailability or errors. The final file considers 37,651 buildings of which 19,311 (51.3%) are properly classified according to the EPC criteria (Table 2). Figure 3 provides a detailed portion of K index energy map at cadastral parcels scale.

Table 2. Summary of resulting data.

Number of buildings included in the vector file	41,010
Number of buildings/parcels considered for the scope	37,651
Number of buildings/parcels related to K index values	19,311

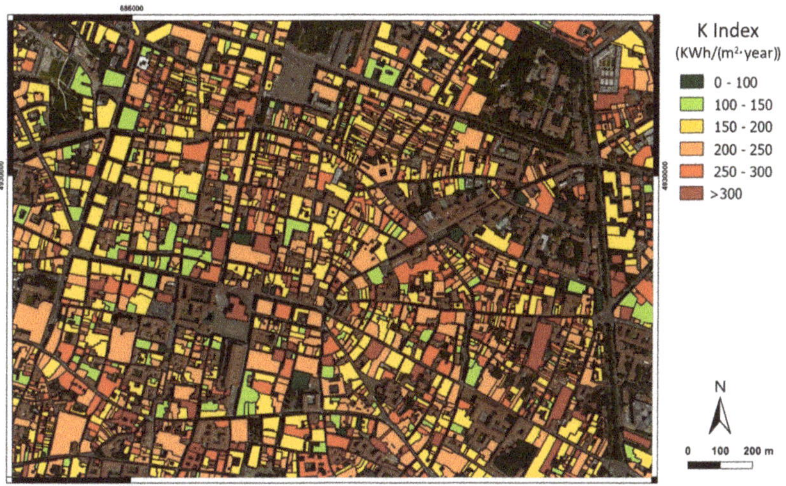

Figure 3. K index map at cadastral parcel scale. Bing aerial imagery is used as the base map.

Figure 4 offers an overview of the K index map at block scale (K$_{BLOCK}$ index) considering the entire extent of Bologna city. Approximately 12% of the blocks (283) resulted as not being classified. The fourth class (orange 200–250 KWh/(m^2·year)) includes the highest number of polygons (960).

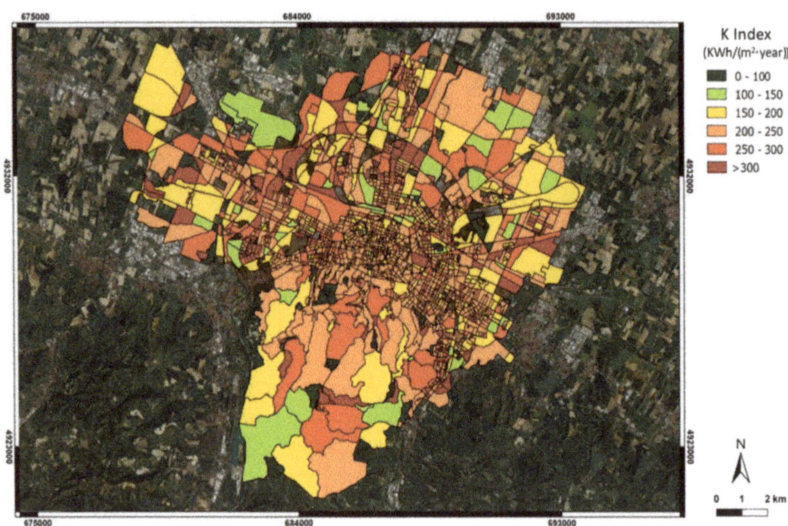

Figure 4. K index map at block scale (K_{BLOCK} index). Bing aerial imagery is used as the base map.

Figure 5 provides the portion of the map at the block scale corresponding with the sample site used for validation purpose. The same color palette used for the K index maps was applied. For each block, the number of EPCs with complete address, available for the Ep_{tot} mean value calculation, is reported.

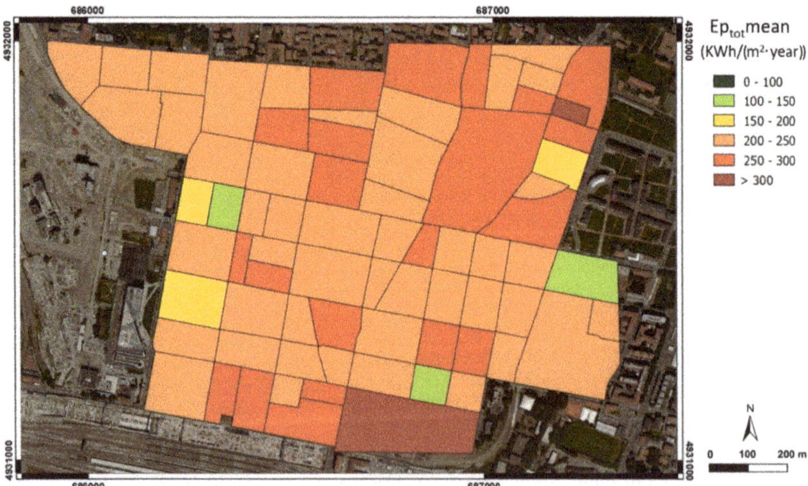

Figure 5. Ep_{tot} mean value map corresponding to the validation area (Bolognina) at block scale. Bing aerial imagery is used as the base map.

Figure 6 provides a detailed map of the K index ($KWh/(m^2 \cdot year)$) in Bolognina district sample site. The blocks requiring further refinement to be classified through validation procedure are highlighted with a cyan line. The K index classification corresponds in 64 out of 80 blocks, while in the other ones the difference never exceeds one class.

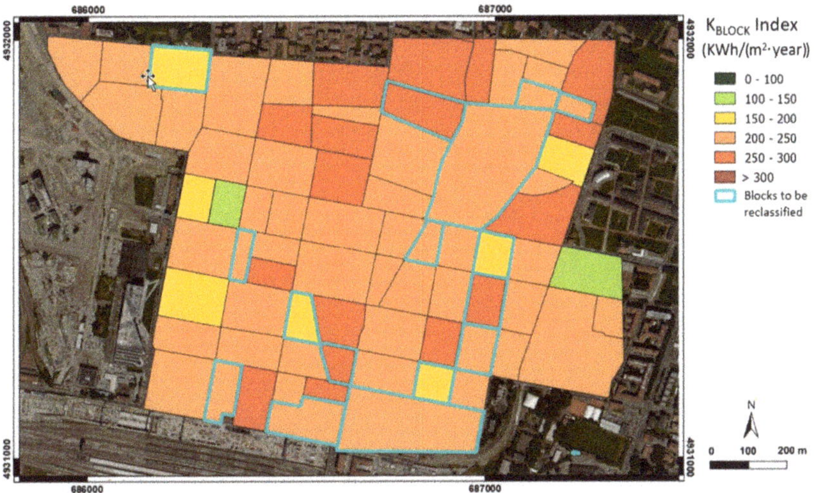

Figure 6. The K_{BLOCK} index map corresponding to the validation area (Bolognina) at block scale. Bing aerial imagery is used as the base map.

6. Discussion

The structure and the stages of the proposed methodological approach clearly reflect the constraints regarding the input data availability, particularly the need to properly set the basic unit on which the study focuses. Within the energy efficiency field, this typically refers to the building scale, which may be coherent with the initial request to obtain a picture of the stock that does not meet the minimum energy efficiency requirements but does not fit well with the scope that instead clearly aims to easily visualize and process data about energy consumption distribution at the district/city scales. The building level is a consolidated basic unit adopted both in GIS and building information modeling (BIM) environments for modeling purposes, as clearly evidenced by the literature [66,67]. However, the limitations of EPCs, due to data protection regulations that require treating data without a full address line (street and street number), obliged us to work at cadastral parcel scale and not at the building scale (the ownership of data may differ according to the context, and in other experiences full data were used according to specific agreements). The complete address and the street number code link each cadastral unit with the corresponding vector geometric element [68].

The MA and the research team took the sensitive data issue seriously from the very beginning, assuming that the methodology has to comply with the Legislative Decree 101, 10/08/2018 [36], that acknowledges the European General Data Protection Regulation 2016/679 [69] within the Italian territory. This is the reason why full address data were made available only for validation purpose on a sample area with limited extent, while the methods developed for using nonsensitive data have been instead applied [70,71], complying with the strict nondisclosure agreement for data processing and considering that the generated outcome would display only aggregated data on the map.

The use of EPCs evidenced some other issues concerning the collection methodology and the data storage organization. EPCs were introduced by the Decree Law n.63 (04/06/2013) [72] and following updates, providing a specific template for all new and existing buildings to be rented or sold within the market. In 2015, some new standards were included in the template in order to make the energy performance classification more homogeneous, partially changing the recorded data structure or the data writing formats. Even though this aspect is not part of the present study, it seriously affected the way in which data have to be processed. Some differences were detected in datasets from 2009 to 2015 and those from 2015 to 2018, requiring additional steps for harmonization [73] during the preparation phase for GIS processing. A recurrent detected error regards the address writing format (names and numbers). The mismatch of the address formulation criteria between the EPC files used for validation and the available vector files obstructs a simultaneous analysis of the entire municipal territory. The frequent typo errors made it impossible to apply an automatic correction procedure, and most of the errors had to be manually amended. This was possible for the sample area (Bolognina), considering its limited extent, but could significantly impact the entire process if applied to the whole city.

As already evidenced by Hohmann [74], these problems suggest a reflection on data collection methods and organization within the MA, especially considering the new three-year plan for information technology in the Public Administration aiming to achieve the digital transformation of the Italian PAs in accordance with the European Digital Agenda programming [75].

Among the detected errors within the EPC files, the one regarding the lack of cadastral data or addresses can be very critical causing the loss of records (which in the specific case study was limited to 5%) and related visualization in the energy map. The lack of geometric congruence between some vector files (such as between building vector and block boundaries vector) can be instead solved by the application of some spatial GIS tools [76].

Taking into account all these limitations, the developed methodology tried to find adequate solutions to generate the energy maps: the K index map was initially obtained at the building scale and then obtained at the cadastral parcel scale, with more than 50% of total geometric units classified.

The number of EPCs used to calculate the K index (referring to the individual cadastral parcel) may vary in each cadastral parcel, consequently the K index of a cadastral parcel can be calculated only as a part of all of the urban units or a part of all buildings included within it. The progressive increase of EPC availability will definitely improve the map reliability until a total coverage is finally achieved. Considering that the available EPCs (covering the time period up to 2018) represent 50% of the urban units (assuming a relevant part of the residual 50% belongs to a building typology that does not require an EPC, as already explained), the developed methodology could be positively refined and implemented during the coming years in order to increase its levels of precision and efficiency.

If the results are observed at the block scale, the K index improves in terms of quality: 88% of the blocks have an associated K-value. Since this scale is more aligned with the planning purposes discussed with the PA, the validation process was performed at the block scale: 80% of the blocks have an associated class corresponding to the K index classification range. Regarding the gap between the two results (conventional methodology and validation), the following considerations can be reported:

- Despite both the conventional and the validation processes using EPCs as input data, the validation does not consider the building step, instead only considering the cadastral parcels in a single step.
- The EPC dataset made available during the validation included 20,000 more records (referring to part of 2018 and 2019), which produced some differences as highlighted in Figures 5 and 6.
- The limited extent of the sample area during the validation process allowed to manually amend some detected errors which otherwise would become a very time-consuming activity if the whole city territory would be involved.

Based on the overall outcomes and these observations, the achieved result was considered to be satisfactory by the MA and the team with relation to the initial research brief.

7. Conclusions

Considering all the premises as well as the limitations and barriers that arose during the study with relation to input data, it has to be honestly said that the outcome confirms the constraints and difficulties already mentioned by the scientific literature. However, the result can be considered a step forward in shifting the perspective from the individual building energy performance to a wider urban energy consumption observation, which was the main requirement of the regional regulation update. This will allow the MA to focus on specific measures where energy intensity concentration requires greater attention or to customize solutions with relation to specific situations. It is also very relevant for launching initiatives (such as incentives or bonus) able to boost a diffuse improvement linked to specific typologies. Much more relevantly, it will be very helpful in monitoring the overall situation detecting where lack of data and information are concentrated as well as in obtaining updated maps which will provide easy-to-visualize feedback of the impact derived by the decisions and energy policies adopted over a time period.

The methodological approach used for generating the K index map can be considered a valid and effective starting point for further detailed elaboration of energy performance maps, taking into account data reliability as an open issue to be further investigated. This can be considered a feasible and realistic goal, assuming that the continuous update of EPCs will make new input data available to feed the methodology, improving its precision and reliability. This will help in taking into account the evolution of energy consumption of buildings over time, keeping the focus on the overall demand and distribution at the city level and supporting reflections on the influence of local and external conditions (such as the effects of climate change). The main benefit deriving from this study lies in the opportunity of using maps to represent a systemic and easy-to-visualize picture of energy consumption at the city level that policy-makers and designers can use to support the decision-making process as well as to easily communicate the outcomes to citizens and end-users, improving the collective level of awareness towards this topic.

It also has to be noted that the described methodology can be generalized and easily replicated in other contexts without its main structure changing, even if the outcomes can be of course affected by the quality of available data and resources. With that being said, this methodology represents an opportunity to explore the energy consumption trends at the city/district scales and consequently address future measures and initiatives to foster energy savings.

Author Contributions: Conceptualization, J.G.; methodology, J.G., E.A., M.D.G.; software, M.D.G.; validation, J.G., E.A., M.D.G.; formal analysis, J.G., E.A., M.D.G.; investigation, J.G., E.A., M.D.G., V.V.; resources, J.G., E.A., M.D.G., V.V.; data curation, M.D.G.; writing—original draft preparation, M.D.G.; writing—review and editing, J.G. and E.A.; visualization, M.D.G.; supervision, J.G. and E.A.; project administration, J.G. All authors have read and agreed to the published version of the manuscript.

Funding: This research was partially funded by Emilia-Romagna Region, within the framework research program titled "Efficienza energetica in edilizia e nel settore industriale" (G.R. 17147/2017, G.R. 19429/2017) supported by the "Piano triennale integrato Fondo Sociale Europeo (FSE), Fondo Europeo di Sviluppo Regionale (FESR) e Fondo Europeo Agricolo per lo Sviluppo Rurale (FEASR) - Alte Competenze per la ricerca, il trasferimento tecnologico e l'imprenditorialità" funding scheme.

Acknowledgments: The authors express their sincere gratitude to V. Orioli, F. Evangelisti, G. Fini, F. Tutino, C. Girotti, C. Manaresi and those who gave their support and their contribution within the Municipality of Bologna.

Conflicts of Interest: The authors declare no conflict of interest.

References

1. Directive 2012/27/EU of the European Parliament and of the Council of 25 October 2012 on Energy Efficiency, Amending Directives 2009/125/EC and 2010/30/EU and Repealing Directives 2004/8/EC and 2006/32/EC. Available online: https://eur-lex.europa.eu/legal-content/EN/TXT/?uri=celex%3A32012L0027 (accessed on 31 October 2019).

2. Directive 2010/31/EU of the European Parliament and of the Council of 19 May 2010 on the Energy Performance of Buildings (Recast). Available online: https://eur-lex.europa.eu/legal-content/EN/TXT/?uri=CELEX%3A32010L0031 (accessed on 31 October 2019).

3. Bhargavi, P.; Jyothi, S. Soil Classification Using Data Mining Techniques: A Comparative Study. *Int. J. Eng. Trends Technol.* **2011**, *2*, 55–59.

4. Cai, G. Contextualization of Geospatial Database Semantics for Human–GIS Interaction. *Geoinformatica* **2007**, *11*, 217–237. [CrossRef]

5. Fonseca, M.E. Using ontologies for integrated geographic information systems. *Trans. Gis.* **2002**, *6*, 231–257. [CrossRef]

6. Jones, C.B.; Alani, H.; Tudhope, D. *Geographical Information Retrieval with Ontologies of Place*; COSIT 2011; Montello, D.R., Ed.; LNCS: Berlin/Heidelberg, Germany, 2001; Volume 2205, pp. 322–335.

7. Lee, A.J.T.; Chen, Y.A.; Ip, W.C. Mining frequent trajectory patterns in spatial–temporal databases. *Inform. Sci.* **2009**, *179*, 2218–2231. [CrossRef]

8. Ravikumar, G.; Sivareddy, M. An Effective Analysis of Spatial Data Mining Methods Using Range Queries. *J. Glob. Res. Comput. Sci.* **2012**, *3*, 7–12.

9. Van Cleef, E. *Cities in Action*; Pergamon Press: New York, NY, USA, 1970.

10. Fischel, W.A. The Evolution of Zoning Since the 1980s: The Persistence of Localism. In *Property in Land and Other Resources*; Cole, D.H., Ostrom, E., Eds.; Lincoln Institute of Land Policy: Cambridge, MA, USA, 2010; pp. 288–293.

11. Erickson, A. The Birth of Zoning Codes, a History. 2012. Available online: https://www.citylab.com/equity/2012/06/birth-zoning-codes-history/2275/ (accessed on 15 April 2020).

12. Trombadore, A. MEETHINK Energy Project A research experience. *Mediterr. Smart Cities* **2016**, *1*, 85.

13. Caputo, P.; Costa, G.; Ferrari, S. A supporting method for defining energy strategies in the building sector at urban scale. *Energy Policy* **2013**, *55*, 261–270. [CrossRef]

14. Theodoridou, I.; Karteris, M.; Mallinis, G.; Papadopoulos, A.M.; Hegger, M. Assessment of retrofitting measures and solar systems' potential in urban areas using Geographical Information Systems: Application to a Mediterranean city. *Renew. Sustain. Energy Rev.* **2012**, *16*, 6239–6261. [CrossRef]

15. Heiple, S.; Sailor, D.J. Using building energy simulation and geospatial modeling techniques to determine high resolution building sector energy consumption profiles. *Energy Build* **2008**, *40*, 1426–1436. [CrossRef]

16. Hill, L. *Georeferencing: The Geographic Associations of Information*; MIT Press: Cambridge, MA, USA, 2009.

17. Zheng, Y.; Zha, Z.; Chua, T. Research and Applications on Georeferenced Multimedia: A Survey. *Multimed. Tools Appl.* **2011**, *51*, 77–98. [CrossRef]

18. Hackeloeer, A.; Klasing, K.; Krisp, J.M.; Meng, L. Georeferencing: A review of methods and applications. *Ann. GIS* **2014**, *20*, 61–69. [CrossRef]

19. Leidner, J. Georeferencing: From texts to maps. *Int. Encycl. Geogr.* **2016**, 1–10. [CrossRef]

20. Reiter, S.; Wallemacq, V. City energy management: A case study on the area of Liège in Belgium. In *Proceedings of the International Conference GEOProcessing 2011*; Rückemann, C.P., Wolfson, O., Eds.; IARIA: Wilmington, NC, USA, 2011; pp. 1–6.

21. Balta, C. GIS-Based Energy Consumption Mapping. Master's Thesis, Edinburgh Research Archive, Edinburgh, UK, 2014.

22. Voigt, P.; Von dem Bussche, A. *The EU General Data Protection Regulation (GDPR). A Practical Guide*, 1st ed.; Springer International Publishing: Cham, Switzerland, 2017.

23. Resch, B.; Sagl, G.; Törnros, T.; Bachmaier, A.; Eggers, J.B.; Herkel, S.; Gündra, H. GIS-based planning and modeling for renewable energy: Challenges and future research avenues. *ISPRS Int. J. Geo. Inf.* **2014**, *3*, 662–692. [CrossRef]

24. Calvert, K. From 'energy geography' to 'energy geographies' Perspectives on a fertile academic borderland. *Prog. Hum. Geogr.* **2016**, *40*, 105–125. [CrossRef]

25. Li, C. GIS for Urban Energy Analysis. In *Comprehensive Geographic Information Systems*; Huang, B., Ed.; Elsevier: Oxford, UK, 2018; pp. 187–195.

26. Parshall, L.; Gurney, K.; Hammer, S.A.; Mendoza, D.; Zhou, Y.; Geethakumar, S. Modeling energy consumption and CO2 emissions at the urban scale: Methodological challenges and insights from the United States. *Energy Policy* **2010**, *38*, 4765–4782. [CrossRef]

27. Van Hoesen, J.; Letendre, S. Evaluating potential renewable energy resources in Poultney, Vermont: A GIS-based approach to supporting rural community energy planning. *Renew. Energy* **2010**, *35*, 2114–2122. [CrossRef]

28. Guttikunda, S.K.; Calori, G. A GIS based emissions inventory at 1 km× 1 km spatial resolution for air pollution analysis in Delhi, India. *Atmos. Environ.* **2013**, *67*, 101–111. [CrossRef]

29. Ascione, F.; Masi, R.F.; De Rossi, F.; Fistola, R.; Sasso, M.; Vanoli, G.P. Analysis and diagnosis of the energy performance of buildings and districts: Methodology, validation and development of Urban Energy Maps. *Cities* **2013**, *35*, 270–283. [CrossRef]

30. Quan, S.J.; Li, Q.; Augenbroe, G.; Brown, J.; Yang, P.P.J. Urban data and building energy modelling: A GIS-based urban building energy modelling system using the urban-EPC engine. In *Planning Support Systems and Smart Cities*; Stan Geertman, S., Ferreira, J., Goodspeed, R., Stillwell, J., Eds.; Springer International Publishing: Cham, Switzerland, 2015; pp. 447–469.

31. Gupta, R.; Gregg, M. Targeting and modelling urban energy retrofits using a city-scale energy mapping approach. *J. Clean. Prod.* **2018**, *174*, 401–412. [CrossRef]
32. Gupta, R.; Gregg, M. Local Energy Mapping Using Publicly Available Data for Urban Energy Retrofit. In *Building Information Modelling, Building Performance, Design and Smart Construction*; Dastbaz, M., Gorse, C., Moncaster, A., Eds.; Springer International Publishing: Cham, Switzerland, 2019; pp. 207–219.
33. Fabbri, K.; Zuppiroli, M.; Ambrogio, K. Heritage buildings and energy performance: Mapping with GIS tools. *Energy Build.* **2012**, *48*, 137–145. [CrossRef]
34. Fabbri, K. Building and fuel poverty, an index to measure fuel poverty: An Italian case study. *Energy* **2015**, *89*, 244–258. [CrossRef]
35. Mattinen, M.K.; Heljo, J.; Vihola, J.; Kurvinen, A.; Lehtoranta, S.; Nissinen, A. Modeling and visualization of residential sector energy consumption and greenhouse gas emissions. *J. Clean. Prod.* **2014**, *81*, 70–80. [CrossRef]
36. Legislative Decree 10/08/2018, n. 101. Disposizioni per L'adeguamento della Normativa Nazionale alle Disposizioni del Regolamento (UE) 2016/679 del Parlamento Europeo e del Consiglio, del 27 Aprile 2016, Relativo alla Protezione delle Persone Fisiche con Riguardo al tRattamento dei Dati Personali, Nonche' alla Libera Circolazione di Tali Dati e Che Abroga la Direttiva 95/46/CE (Regolamento Generale Sulla Protezione Dei Dati). Available online: https://www.gazzettaufficiale.it/eli/id/2018/09/04/18G00129/sg (accessed on 10 November 2019).
37. Kolter, J.Z.; Johnson, M.J. REDD: A public data set for energy disaggregation research. In *Workshop on Data Mining Applications in Sustainability (SIGKDD)*; Citeseer: San Diego, CA, USA, 2011; Volume 25, pp. 59–62.
38. Porse, E.; Derenski, J.; Gustafson, H.; Elizabeth, Z.; Pincetl, S. Structural, geographic, and social factors in urban building energy use: Analysis of aggregated account-level consumption data in a megacity. *Energy Policy* **2016**, *96*, 179–192. [CrossRef]
39. Tronchin, L.; Fabbri, K. Energy Performance Certificate of building and confidence interval in assessment: An Italian case study. *Energy Policy* **2012**, *48*, 176–184. [CrossRef]
40. Murphy, L. The influence of the energy performance certificate: The Dutch case. *Energy Policy* **2014**, *67*, 664–672. [CrossRef]
41. Ballarini, I.; Corgnati, S.P.; Corrado, V. Use of reference buildings to assess the energy saving potentials of the residential building stock: The experience of TABULA project. *Energy Policy* **2014**, *68*, 273–284. [CrossRef]
42. Ballarini, I.; Corgnati, S.P.; Corrado, V.; Talà, N. Definition of building typologies for energy investigations on residential sector by TABULA IEE-project: Application to Italian case studies. *Roomvent Trondheim* **2011**, 19–22.
43. Loga, T.; Stein, B.; Diefenbach, N. TABULA building typologies in 20 European countries—Making energy-related features of residential building stocks comparable. *Energy Build.* **2016**, *132*, 4–12. [CrossRef]
44. Escrivá-Escrivá, G.; Álvarez-Bel, C.; Peñalvo-López, E. New indices to assess building energy efficiency at the use stage. *Energy Build.* **2011**, *43*, 476–484. [CrossRef]
45. Do, H.; Cetin, K.S. Residential Building Energy Consumption: A Review of Energy Data Availability, Characteristics, and Energy Performance Prediction Methods. *Curr. Sustain. Renew. Energy Rep.* **2018**, *5*, 76–85. [CrossRef]
46. Hong, T.; Taylor, L.S.C.; D'Oca, S.; Yan, D.; Corgnati, S.P. Advances in research and applications of energy-related occupant behavior in buildings. *Energe Build.* **2016**, *116*, 694–702. [CrossRef]
47. Cetin, K.S.; Siemann, M.; Sloop, C. Disaggregation and future prediction of monthly residential building energy use data using localized weather data network. *2016 ACEEE Summer Study Energy Effic. Build.* **2016**, *12*, 1–12.
48. Wilde, P. The gap between predicted and measured energy performance of buildings: A framework for investigation. *Autom Constr.* **2014**, *41*, 40–49. [CrossRef]
49. Maantay, J.; Ziegler, J.; Pickles, J. *GIS for the Urban Environment*; Esri Press: Redlands, CA, USA, 2006.
50. ISTAT, Statistiche Demografiche ISTAT. Available online: http://demo.istat.it/bilmens2019gen/index.html (accessed on 31 March 2019).
51. Bernabei, G.; Gresleri, G.; Zagnoni, S. *Bologna Moderna 1860–1980*; Patron: Bologna, Italy, 1984.
52. Alaimo, A. *L'organizzazione della Città: Amministrazione Comunale e Politica Urbana a Bologna Dopo l'Unità (1859-1889)*; Il Mulino: Bologna, Italy, 1990.

53. Bazzoli, N. *Frammentazione Urbana e Nuove Dinamiche Insediative. Bologna e il Suo Hinterland. Sguardi sul Mondo. Percorsi di Geografia Sociale*; Patron: Bologna, Italy, 2014; pp. 215–224.

54. BLEUAP. Bologna Local Urban Environment Adaptation Plan for a Resilient City. Available online: http://www.blueap.eu/site/ (accessed on 15 June 2019).

55. TABULA WebTool. Available online: http://webtool.building-typology.eu/#bm (accessed on 10 April 2019).

56. Gazzetta Ufficiale. Available online: https://www.gazzettaufficiale.it/eli/id/2013/08/03/13G00133/sg (accessed on 31 October 2019).

57. Economidou, M. (Ed.) *Europe's Buildings under the Microscope. A Country-by-Country Review of the Energy Performance of Buildings*; Buildings Performance Institute Europe: Brussels, Belgium, 2011.

58. Braulio-Gonzalo, M.; Bovea, M.D.; Ruá, M.J.; Juan, P. A methodology for predicting the energy performance and indoor thermal comfort of residential stocks on the neighbourhood and city scales. A case study in Spain. *J. Clean. Prod.* **2016**, *139*, 646–665. [CrossRef]

59. Mastrucci, A.; Baume, O.; Stazi, F.; Leopold, U. Estimating energy savings for the residential building stock of an entire city: A GIS-based statistical downscaling approach applied to Rotterdam. *Energy Build.* **2014**, *75*, 358–367. [CrossRef]

60. Droutsa, K.G.; Kontoyiannidis, S.; Dascalaki, E.G.; Balaras, C.A. Mapping the energy performance of hellenic residential buildings from EPC (energy performance certificate) data. *Energy* **2016**, *98*, 284–295. [CrossRef]

61. Liu, Y.; Yang, L.; Zheng, W.X.; Liu, T.; Zhang, X.R.; Liu, J.P. A novel building energy efficiency evaluation index: Establishment of calculation model and application. *Energy Convers. Manag.* **2018**, *166*, 522–533. [CrossRef]

62. Pérez-Lombard, L.; Ortiz, J.; Rodríguez Maestre, I. A review of benchmarking, rating and labelling concepts within the framework of building energy certification schemes. *Energy Build.* **2009**, *41*, 272–278. [CrossRef]

63. Hai, Z.; Frédéric, M. A review on the prediction of building energy consumption. *Renew Sustain. Energy Rev.* **2012**, *16*, 3586–3592.

64. Codd, E.F. A relational model of data for large shared data banks. *Commun. ACM* **1970**, *13*, 377–387. [CrossRef]

65. Bologna Municipality. Available online: http://umap.openstreetmap.fr/it/map/sezioni-di-censimento-bologna_125580#15/44.4954/11.3397 (accessed on 20 February 2020).

66. Ham, Y.; Golparvar-Fard, M. Mapping actual thermal properties to building elements in gbXML-based BIM for reliable building energy performance modelling. *Autom. Constr.* **2015**, *49*, 214–224. [CrossRef]

67. Davila, C.C.; Reinhart, C.F.; Bemis, J.L. Modeling Boston: A workflow for the efficient generation and maintenance of urban building energy models from existing geospatial datasets. *Energy* **2016**, *117*, 237–250. [CrossRef]

68. ESRI. GIS to Meet Renewable Energy Goals. Available online: http://www.esri.com/news/arcnews/fall09articles/gis-to-meet.html (accessed on 3 November 2019).

69. Regulation (EU) 2016/679 of the European Parliament and of the Council of 27 April 2016 on the Protection of Natural Persons with Regard to the Processing of Personal data and on the Free Movement of Such Data, and Repealing DIRECTIVE 95/46/EC (General Data Protection Regulation). Available online: https://eur-lex.europa.eu/eli/reg/2016/679/oj (accessed on 3 November 2019).

70. Pencarrick Hertzman, C.; Meagher, N.; McGrail, K.M. Privacy by Design at Population Data BC: A case study describing the technical, administrative, and physical controls for privacy-sensitive secondary use of personal information for research in the public interest. *J. Am. Med. Inform. Assoc.* **2012**, *20*, 25–28. [CrossRef]

71. Gholami, A.; Laure, E. Security and privacy of sensitive data in cloud computing: A survey of recent developments. *arXiv* **2016**, arXiv:1601.01498. [CrossRef]

72. Legislative Decree 04/06/2013, n. 63. Disposizioni Urgenti per il Recepimento della Direttiva 2010/31/UE del Parlamento Europeo e del Consiglio del 19 Maggio 2010, Sulla Prestazione Energetica nell'edilizia per la Definizione delle Procedure D'infrazione Avviate dalla Commissione Europea, Nonche' Altre disposizioni in Materia di Coesione Sociale. Available online: https://www.gazzettaufficiale.it/eli/id/2013/06/05/13G00107/sg (accessed on 10 November 2019).

73. Nijssen, G.M. *Modelling in Data Base Management Systems*; North-Holland Pubnlishing: Amsterdam, The Netherlands, 1976.

74. Hohmann, L.; Rappapor, I.S.; Schnitz, M.; Rosenquis, B.; Jackson, A. Intellectual Asset Protocol for Defining Data Exchange Rules and Formats for Universal Intellectual Asset Documents, and Systems, methods, and Computer Program Products related to Same. U.S. Patent Application No. 7,437,471, 2 March 2008.

75. AGID (Agenzia per l'Italia Digitale). Available online: https://www.agid.gov.it\T1\guilsinglrightrepository_files\T1\guilsinglrightPiano-Triennale-ICT-2019-2021 (accessed on 3 December 2019).

76. QuantumGis Tutorial. Available online: https://docs.qgis.org/testing/en/docs/training_manual/ (accessed on 3 December 2019).

Article

Assessment of Thermal Comfort in the Intelligent Buildings in View of Providing High Quality Indoor Environment

Grzegorz Majewski [1], Łukasz J. Orman [2,*], Marek Telejko [3], Norbert Radek [4], Jacek Pietraszek [5] and Agata Dudek [6]

[1] District Court, Warszawska 1, 26-610 Radom, Poland; majewskigrzegorz@wp.pl
[2] Faculty of Environmental, Geomatic and Energy Engineering Kielce University of Technology, 25-314 Kielce, Poland
[3] Faculty of Civil Engineering and Architecture, Kielce University of Technology, 25-314 Kielce, Poland; mtelejko@tu.kielce.pl
[4] Mechatronics and Mechanical Engineering, Kielce University of Technology, 25-314 Kielce, Poland; norrad@tu.kielce.pl
[5] Faculty of Mechanical Engineering, Cracow University of Technology, 31-864 Cracow, Poland; jacek.pietraszek@mech.pk.edu.pl
[6] Institute of Materials Engineering, Czestochowa University of Technology, 42-200 Częstochowa, Poland; dudek@wip.pcz.pl
* Correspondence: orman@tu.kielce.pl

Received: 31 March 2020; Accepted: 15 April 2020; Published: 16 April 2020

Abstract: The paper analyses the indoor environment in two modern intelligent buildings located in Poland. Measurements of air and globe temperatures, relative humidity and carbon dioxide concentration in 117 rooms carried out in the space of 1.5 years were presented. Thermal comfort of the occupants has been investigated using a questionnaire survey. Based on 1369 questionnaires, thermal sensation, acceptability and preference votes were analysed in view of their interdependency as well as their dependency on operative temperature, which proved to be very strong. It has been found that the respondents did not completely rate thermal comfort and indoor environment quality as very high, although the overwhelming sensations were positive. Apart from the operation of heating, ventilation and air conditioning (HVAC) systems, this might have also been the cause of individual human factors, such as body mass index, as tested in the study, or the finding that people were generally in favour of a warmer environment. Moreover, thermal environment proved to be the most important element for ensuring the well-being of the occupants.

Keywords: indoor environment; intelligent buildings; thermal comfort; thermal sensation

---

1. Introduction

Energy consumption throughout the world constantly rises. Much of it is related to the use in the building sector, especially for the development and daily operation of sophisticated modern buildings such as large-scale intelligent buildings for office and public use. Those types of buildings are equipped with services and installations such as heating, ventilation and air conditioning (HVAC) systems, whose energy consumption might be significant. In commercial facilities the main electrical loads are typically related to HVAC systems and lighting [1], in office buildings heating, ventilation and air conditioning systems may account for over 60% of energy consumption [2], while data from three years of final energy consumption in a modern educational building in France indicate over 80% energy use for heating and ventilation [3]. Thus, even a small reduction in indoor air temperature during the heating season and a small increase in indoor temperature during the summer (reducing the required cooling load) might result in both economic and environmental benefits due to reduced operational costs and lower emission of carbon dioxide if energy is generated from fossil fuels. The only question is to which extend the temperatures could be reduced/increased, so that the residents would still experience thermal comfort conditions.

Modern buildings have been considered and designed with the view to provide their occupants with comfort conditions—both in terms of indoor air quality and thermal comfort. However, these requirements might not be satisfied and people complain about the indoor air condition despite the high investment and operational costs of the buildings. Moreover, the long-term policy and development strategy of the European Union states that buildings in member states should be characterised, among other things, by reduced energy consumption and improved indoor air quality [4]. In view of these reasons, much more effort needs to be made to target the problems currently prevalent in the area of civil engineering such as sick building syndrome and, in particular, provide more knowledge on thermal comfort and indoor air quality in modern buildings.

Thermal comfort can be defined as the feeling of neither cold nor warm. It is affected by a number of parameters such as indoor air and radiant temperatures, relative humidity, air movement as well clothing thermal resistance and metabolism rate. They are incorporated in the Fanger model, which predicts collective occupants' satisfaction taking into account six environmental and physiological parameters [5]. The author proposed the Predicted Mean Vote (PMV) and Percentage of People Dissatisfied (PPD) indices, which are now part of the commonly used building standards: EN-ISO 7730 [6] and ASHRAE-55 [7]. The PMV is a seven-point scale ranging from cold (−3) to hot (+3) and describes the thermal sensation of a large number of people in the room. The generally preferred value is 0 (neutral thermal environment) with deviations into the positive and negative area, whose acceptable level depends on the building category and is given in the standard [8].

Thermal comfort has recently attracted a lot of scientific interest, as evidenced by a growing number of experimental papers and reviews [9,10]. This phenomenon is tested in all types of buildings such as residential [11], public utility buildings [12,13] or even means of transport [14]. In terms of public utility buildings, where people tend to spend more and more time and are less able to control the indoor environment than at their homes, educational facilities and various kinds of offices should be of particular interest both in terms of ensuring optimal thermal sensation and reducing heating/cooling costs of HVAC systems (for example a rise in indoor temperature set-point of ca. 1–2 °C in the summer can save ca. 6–10% of the electric energy for air conditioning [15]). Moreover, much attention has recently been paid to combining the issue of providing thermal comfort conditions to room users and the energy performance of buildings [16–18].

Almeida et al. [19] performed 32 measurements in ten spaces in six educational buildings in Portugal and obtained 490 questionnaires. PMV index was determined for each room. Air and mean radiant temperatures as well as air velocity and relative humidity were recorded using a measuring device. The questionnaire covered questions regarding thermal perception, thermal evaluation, thermal preference and thermal tolerance. High acceptability of the indoor conditions was observed. Over half of the respondents intended to keep the thermal environment unchanged. The relationship between PMV and mean thermal sensation vs. operative temperature turned out to have a moderate correlation. Krawczyk et al. [20] presented research results of temperature and relative humidity measurements in Polish and Spanish universities as well as surveys on the comfort perception. It was found that the optimal indoor air temperature differs depending on location (for students in Poland range from 21.7 to 22.3 °C, while in Spain from 23.3 to 24.8 °C). However, both indoor air temperature and humidity in both buildings are within the recommended range. Most Spanish students felt that is was cold or too cold for them, despite the fact that the temperature was within the recommended range and ca. 2 °C higher than in Poland (which was explained by adaptive models for the comfort temperature (higher values for Spain than for Poland for this month). The respondents proved to have good acceptability of dry conditions. Vilcekova et al. [21] carried out tests of indoor air quality and thermal comfort in five classrooms located in a school in Slovakia. The following parameters were measured: temperature, relative humidity, solid particles, CO_2 concentration, sound level as well as lighting conditions. The number of people tested was 34 students of 6–15 years old and five teachers. Thermal sensations were overwhelmingly acceptable and perceived as good. A concern was the relatively high CO_2 level (in the range from 577 to 1787 ppm) despite the fact that windows were open. The concentration of carbon dioxide fluctuated during the school hours. Fang et al. [22] analysed thermal comfort in air-conditioned classrooms located in Hong Kong. The research programme consisted of measurements of environmental parameters as well as thermal responses from 982 students. The authors found out that there was an insignificant difference between the temperature preferred by the respondents and the neutral temperature. Besides, there was a strong linear relation between the mean thermal sensation vote and the operative temperature. Buratti and Ricciardi [23] used direct measurements of the indoor microclimate parameters and the questionnaires during autumn, winter and spring in classrooms at the University of Perugia, Terni and Pavia in Italy. The internal air temperatures ranged from 20.9 to 26.4 °C, mean radiant temperatures from 20.1 to 25.1 °C, while relative humidity from 25% to 63.7%. Air movement was low, with maximal air velocity of 0.085 m/s. Thermal sensations experienced by the respondents in the considered classrooms proved to be very satisfactory (mostly falling within the acceptable range of −0.5–0.5, except for two surveys that produced the results of thermal sensation vote of 0.6 and 0.78). A later work by these authors [24] provided more diversified results. It also focused on other parameters affecting the human well-being, namely: acoustic and lightning conditions. The tests were conducted in classrooms at the University of Pavia in Italy. The tests covered measurements of physical parameters (indoor and external air temperature and relative humidity, mean radiant temperature and air velocity) and questionnaires for respondents' responses (whose total number was 331). It was observed that the average values of PMV and PPD for women and men can differ from each other (maximally by 0.45% and 11.57%, respectively). Generally, thermal comfort could be considered as very good for three classrooms (mean thermal sensation vote ranged from 0.08 to 0.36) and unsatisfactory for two (where mean sensation vote reached 1.15 and −1.10). Aghniaey et al. [25] provided data on thermal comfort at the University of Georgia'a campus in US. The results were obtained in 11 classrooms where the number of students ranged from 18 to 54 persons. The operative temperature was between 21 and 27 °C. The study showed that the thermal environment in the analysed classrooms was overwhelmingly acceptable. The average thermal preference vote was "no change" if the indoor operative temperature was ca. 23.5 °C (thus, this temperature value can be considered optimal at the considered location). For this temperature the actual mean vote was almost neutral. Women typically preferred warmer thermal environments and on average female thermal sensation votes were lower than those for men.

It needs to be noted that the optimal human thermal conditions are subjective and depend on the climate. Experimental results on thermal comfort tests conducted in the summer season in India [26] reveal that the mean comfort temperature (determined by Griffiths' method) was 29.8 °C. In this paper thermal comfort of 900 students (mostly males) in naturally ventilated classrooms during the summer in India was analysed. The study covered thermal sensations, preferences and acceptability as well as measurements of environmental parameters. The mean indoor air temperature was 30.4 °C (from the range of ca. 26.6 to ca. 36 °C), while mean relative humidity was 39.4% and clothing thermal resistance ranged from ca. 0.41% to ca. 59% of votes which indicated the neutral thermal sensation and 69% chose "no change" as their thermal preference. About 81% of the responses were in the comfort band. Thus, the research shows that the overwhelming majority was satisfied with the conditions that might be considered unsatisfactory for the European respondents.

One of the relatively few works on thermal comfort in intelligent educational buildings has been presented by Merabtine et al. [3]. The authors investigated indoor air quality and thermal comfort in a low energy consumption building in France. The authors tested 41 students between the ages of 17 and 22 years using questionnaires in the foyer of the building. The indoor air parameters were measured. The obtained results indicate a dissatisfaction with the thermal conditions in the considered space (beyond the 10% acceptance level of PPD). Thermal sensation votes were also outside of the acceptable range and fell below −1. When the occupancy in the building was high the CO_2 levels exceeded the limit of 820 ppm, but it was rare (1–5% of occupancy hours). In the tested three teacher rooms, the indoor air temperature ranged from ca. 20.3 to 24.7 °C, while CO_2 concentration from ca. over 300 up to over 1300 ppm. Based on these data, it can be concluded that even such modern buildings might not offer comfortable conditions to their users. Carbon dioxide concentration proved to exceed the upper limit at times, while thermal sensation votes of the respondents indicated the feeling of dissatisfaction.

Apart from schools and university campuses, office buildings are also locations where people spend a great deal of time, either as employees or clients. Thus, the energy performance and thermal comfort of the occupants are a crucial issue.

Moujalled et al. [2] conducted thermal comfort measurement in five naturally ventilated office buildings in the south of France. The indoor air temperatures ranged from 21.1 to 32.9 °C. In total 120 people took part in the survey, generating 374 data sets. It was concluded that during the warm season over half of the respondents were dissatisfied with the indoor microclimate. During the cold season over 90% were comfortable. The thermal sensation votes proved to be well correlated with the operative temperature. The room users turned out to be less sensitive to an increase in temperature during the warm season.

Hens [27] performed questionnaire and direct measurement surveys in two air-conditioned office buildings in Belgium. It was reported that the number of the dissatisfied at the mean vote of 0 was greater than 5% (as given in the standard based on the Fanger model). A relatively high number of health complaints was reported by the occupants, who suffered from headache (ca. 44% of respondents), dry throat or blocked nose (over 30%). This might influence the subjective thermal sensation votes expressed by the people.

Kuchen and Fisch [28] conducted tests in 148 work spaces in 25 office buildings at different locations in Germany. The total number of measurements was 345 and all took place during winter (not older than ten years). The most preferred operative temperature value was within the range of 21–22 °C (here, the average thermal sensation vote was 0.08 and 0.12, respectively). The authors observed a relatively good linear relation between the operative temperature and neutral temperature.

Ricciardi and Buratti [29] experimentally analysed thermal comfort in nine open space offices in Italy. Indoor environmental parameters were measured and 588 questionnaires were statistically considered. Indoor air temperature ranged from 22.1 to 26.2 °C. Thermal sensation proved to be generally fine with PMV values ranging from −1.38 to +2.00 and a high mean thermal dissatisfaction of 60% was observed. The correlation between operative temperature vs. PMV was considered good and between neutral and operative temperatures relatively good.

Indraganti et al. [30] conducted tests in four office buildings in Japan during the summer. Four environmental as well as two personal variables were measured together with CO_2 concentration, while 435 occupants were surveyed (which produced 2402 questionnaires). The mean indoor air temperature was 28.2 °C, while mean CO_2 concentration 1055 ppm. The comfort temperature turned out to be 27.1 °C. The majority of people voted in the comfort band of −1 to +1, thus thermal acceptability from the questionnaires proved to be high. The authors observed that the mean sensation vote in buildings with natural ventilation was always higher than in air conditioned buildings. This finding was partly confirmed in another paper by the same authors [31], who performed tests in 25 office buildings in India. However, in this case—having analysed 2612 questionnaires from 1658 volunteers—they found that thermal acceptability was generally low (62.5%), especially in buildings with natural ventilation.

The results of the experimental work of Jazizadeh et al. [32] on thermal comfort in office buildings reveal that ambient temperature is the most important parameter in terms of the thermal sensation of users. However, other factors (not often investigated) such as carbon dioxide concentration and light intensity also play a role—in some rooms their impact in smaller and in some larger in comparison to humidity. Szabo and Kajtar [33] investigated the influence on the type HVAC system on thermal comfort in three office buildings. The best results were obtained for the ducted fan-coil with suspended ceiling installation. The indoor air quality measured with the CO_2 concentration proved to be high in all the buildings. Fedorczak-Cisak et al. [34] considered the impact of external shading equipment in office rooms with large glazed areas on indoor thermal comfort. It was observed that the application of external Venetian blinds reduced the ambient air temperatures and the number of discomfort hours by 92%. However, the tests did not cover the actual thermal sensation measurements based on human responses—only calculated results of PMV and PPD were provided according to the standard.

Due to the fact that heating, ventilation and air conditioning systems in buildings consume much thermal and electrical energy, any reduction in heating during the cold season (linked with decreased indoor temperatures) and in cooling requirements in summer (leading to elevated indoor temperatures) may provide significant economic and environmental savings. However, such adjustments should not have an unfavourable impact on the thermal comfort of the occupants. Thus, more information is needed as to the acceptable levels of indoor microclimate parameters. Moreover, the indoor environment might have an impact on work productivity [35] and learning performance [36,37]. Thus, the issue of ensuring thermal comfort conditions needs to be considered already at an early design stage of public utility buildings.

The paper analyses thermal comfort and general sensation experienced by people in two intelligent buildings located in Poland, seventy kilometers apart from each other. Based on the literature review, it needs to be noted that such a comprehensive study focused on the eastern European region, as presented in this paper, has not been found in literature.

2. Materials and Methods

The analysis covered two modern intelligent buildings in Poland. The first one is an educational facility "Energis" housing the Faculty of Environmental, Geomatic and Energy Engineering of Kielce University of Technology (Figure 1), while the second is the courthouse in Radom (Figure 2).

Figure 1. "Energis" building of Kielce University of Technology (Western façade).

Figure 2. Courthouse in Radom (Northern façade).

Definitions of "intelligent buildings" have changed and evolved over time as pointed out in the review paper [38], however according to the classic definition proposed by Leifer [39] the fundamental constituents of such buildings are, among others: automation, building management systems, information communication networks.

Both the considered buildings are equipped with heating, ventilation and air conditioning as well as building management systems. The air processing units filter the outside air and preheat it (if necessary) before it is supplied via the ductwork to individual rooms. In those units, moisture and heat are recovered with the rotary heat exchangers. However, radiators and floor heating (only in the Energis building) are the main sources of heat. Cooling is provided with fan-coil devices in the courthouse and with 4-way ceiling-mounted indoor units in Energis. The cooling liquid (solution of ethylene glycol) is cooled with chillers located on the rooftops and circulated to indoor units if there is such a need. Each room is equipped with temperature control panel that enables occupants to set the required temperature value. However, this parameter can also be set by the building operator using the building management system (BMS) interface. This system also enables full monitoring of the buildings' parameters, including energy consumption and generation (only in Energis, where electricity is produced from photovoltaic panels, while heat is provided using heat pumps and solar collectors), lighting, access to rooms as well as weather data. Apart from that, it integrates visual anti-theft monitoring, fire protection systems and other functions. BMS can be accessed via internet and the operators can maintain constant remote control over the buildings. Thus, such a complete integration of the HVAC and monitoring systems with BMS as well as the use of automation and control in order to provide the occupants with the adequate indoor conditions can be considered as building "intelligence". Individual occupants of rooms can also consider the building as intelligent, because it can maintain the pre-set indoor conditions despite changes occurring outdoors or indoors.

Both these buildings are fairly new and are located about seventy kilometers from one another. The Energis building was completed in the year 2012, while the courthouse in 2017. Table 1 summarises their main technical data.

Table 1. The main technical data of the analysed intelligent buildings.

No.	Parameter	Energis Building in Kielce	Courthouse in Radom
1	Building area	925 m^2	3239.30 m^2
2	Total floor area	6288.92 m^2	14,822.48 m^2
3	Usable floor area	5121.24 m^2	5640.17 m^2
4	Cubature	21,211 m^3	61,661.48 m^3
5	External wall (construction material/thickness)	Ceramic hollow trick/25 cm, 30 cm	Reinforced concrete/15 cm, 25 cm, 30 cm
6	U value of the external wall	0.17 W/(m^{2}K)	0.28 W/(m^{2}K)

The buildings have been properly insulated to ensure low heat losses to the environment, which is particularly important in Poland, where winters can be severe with outside air temperatures falling to ca. −25 °C. The thermographic analysis conduced in both buildings revealed no major problems or defects. Figures 3 and 4 present the investigation results of thermal fields observed from the outside and in selected rooms.

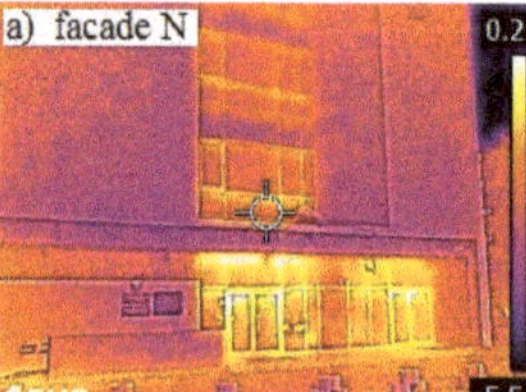

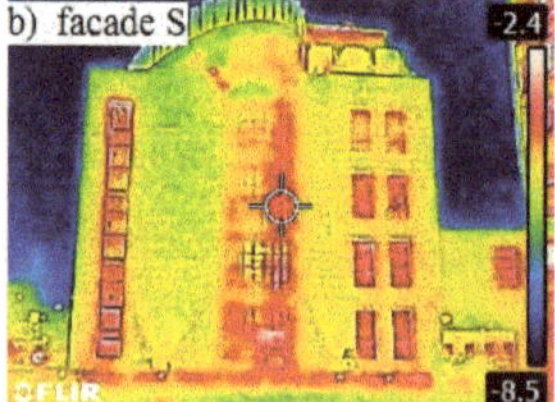

Figure 3. *Cont.*

Figure 3. Energis: thermal images of the Northern (**a**) and Southern (**b**) façades as well as lecture rooms on the fourth (**c**) and first floor (**d**).

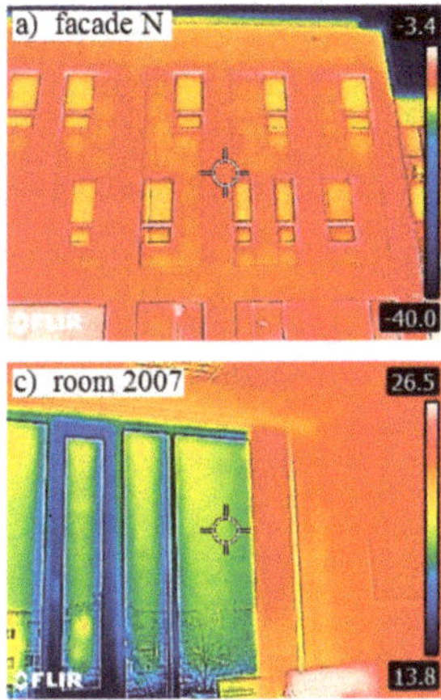
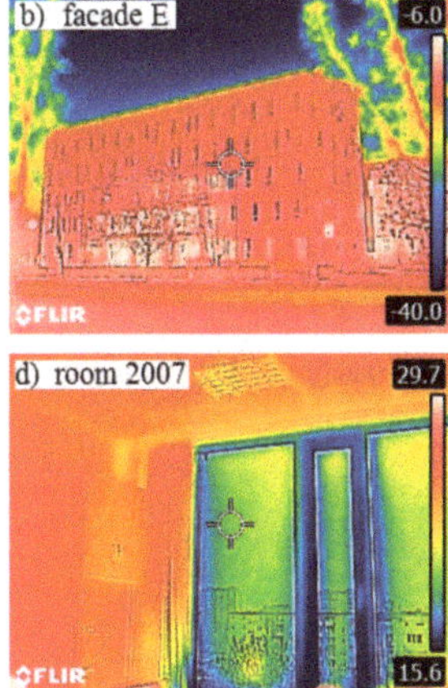

Figure 4. The courthouse: thermal images of the Northern (**a**) and Eastern (**b**) façades as well as room 2007 on the second floor (**c,d**).

It needs to be noted that Energis is close to the passive building standard. It utilises a number of renewable energy sources such as ground and air source heat pumps, solar panels as well as photovoltaic systems.

The tests were conducted in the space of 1.5 years, between February 2017 and July 2018, which enabled us to cover all the seasons. They consisted of simultaneous measurement of environmental physical parameters in rooms, namely: air temperature, globe temperature, air velocity, relative humidity and carbon dioxide concentration. The tests have been performed with two devices, first of all with BABUC-A data logger (by the Italian company Lsi-Lastem) as well as a microclimate meter Sensotron PS32 (by the Polish company Sensotron). Additionally, the infrared images were obtained with FLIR E5 thermovision camera in order to determine possible presence of thermal bridges. Table 2 presents the details of the testing equipment.

Table 2. Technical details of the measuring system.

Number	Parameter	Measuring Range	Resolution	Accuracy
1	Air temperature	10–45 °C	0.1 °C	0.5 °C
2	Relative humidity	0–100%	0.1%	3%
3	Globe temperature	−50–70 °C	0.1 °C	1 °C
4	CO_2 concentration	0–5000 ppm	1 ppm	20 ppm + 3% of the measured value
5	Air velocity	0–30 m/s	0.01 m/s	3 % (at 25 °C)

The measuring device was located on the tripod in the central part of the analysed rooms. Figure 5a and b present the experimental set-up in a lecture room in the Energis building and in an office in the courthouse, respectively.

Figure 5. Experimental set-up: microclimate meter in the lecture room in Energis building (**a**) and in the office room in the courthouse (**b**).

The tests were carried out in 47 different rooms in both buildings. The largest number of people in one room (a lecture theater in the Energis building) was 65. The data obtained in the course of the experiment were recorded and stored for further processing. At the same time as the measurements were taken, the occupants of a given room were provided with questionnaires. These volunteers were asked to answer 36 questions regarding their thermal sensation, preference and acceptability as well as humidity, air quality, activity level before coming to the room, and clothing etc. In total, 118 measurements were taken in different rooms (some were used more than once)—77 comprised of rooms with at least 5 people while 41 in rooms of fewer than 5 users. The total number of questionnaires was 1501, while 1369 of these related to rooms containing at least 5 people. The database consisted of votes from healthy people only (questionnaires, in which the respondent expressed that he/she was ill were not considered). It results from the fact that thermal sensation of people who are unwell might not be objective. The measurements and the questionnaire survey were performed after at least 30 min of the occupants entering the room. This time is enough to adopt to the new environment, as pointed out by Mishra et al. [40].

The monitoring activities in each room lasted up to 1–1.2 h. Upon switching on, the measuring system required a certain amount of time to stabilise and the respondents needed to accommodate to the environment and its conditions (up to 30–40 min). After that time the volunteers were asked to fill in the questionnaires. The sampling rate was set at 1 min. The measurement results of the physical parameters taken during and shortly before the respondents filled in the questionnaires indicate that the levels of all the parameters, except for the carbon dioxide concentration, were quite stable (with standard deviation values not exceeding the measurement error, as given in Table 2). In terms of the CO_2 concentration, it is significantly dependent on the number of people in each room. Generally, room occupancy was high for the office rooms in the courthouse and varied considerably during lectures in the Energis building (Table S1 in the Supplementary material provides details on room occupancy, while Table S2 shows carbon dioxide concentration). Generally, almost all the lecture rooms and classrooms in the building were occupied at the time of measurements. However, sometimes the attendance was low and/or a relatively small group occupied a large room.

The largest number of respondents were full-time students in the Energis building. Their age typically ranged from 19 to 24 years old, while the age of office workers employed at the courthouse varied significantly and was up to 63 years old. Generally, the average age of all the respondents was 25.1 (standard deviation 8.5), while the median age was 22. The body mass index (BMI) ranged from

15.1 to 43.9, with a mean value of 22.8 (standard deviation 3.5) and a median of 22.2. Women made up 55.7% of the population in the survey, while men accounted for 44.3%.

3. Results and Discussion

3.1. Indoor Air Parameters in the Intelligent Buildings

Both Energis and Courthouse intelligent buildings have been designed and built to ensure proper indoor air conditions for people, who are currently there. However, despite much investment cost, the indoor air parameters determined in the course of the experimental programme sometimes proved to be less then satisfactory. Figure 6 presents the test results of relative humidity and operative temperature for 118 rooms (some tested more than once). The operative temperature has been calculated as the arithmetic mean of indoor air and mean radiant temperatures (as in [25]) due to the low values of air velocity.

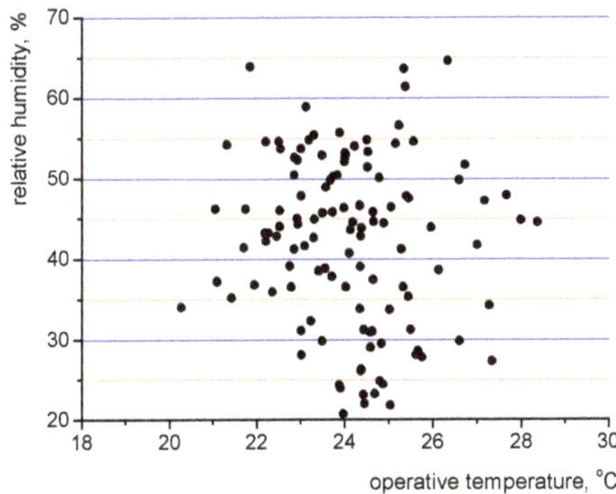

Figure 6. Operative temperature and relative humidity values in selected rooms between February 2017 and July 2018.

For some rooms the test results indicate that the indoor microclimate can be considered unsatisfactory. Operative temperature exceeds the maximal value of 26 °C as given in the standard [8] for eleven measurement sets (ca. 9% of cases). Although, it needs to be noted that this might have happened due to the preference of the room occupants, who might have liked such warm environments and were unwilling to switch on air cooling. In terms of relative humidity, its optimal value depends on a number of parameters, however some data points in Figure 6 fall below the 25% level (almost 7% of cases). Consequently, a sensation of discomfort might be generated, if solid particles or other air contaminants are present in the air.

During the surveys carbon dioxide concentration was also determined. Its required value, like in the case of operative temperature, depends on the building category. For the buildings within category II (to which both analysed buildings belong) the upper limit is 800 ppm plus the outside air CO_2 level [8]. Figure 7 presents the data of indoor air temperature and carbon dioxide concentration from 118 measurement sets. The analysis of the figure indicate that some measurements are characterised by elevated levels of carbon dioxide concentration.

Undoubtedly, the indoor air parameters can be improved in order to provide more comfort to the occupants. Another issue is the subjective assessment and responses of room users to such indoor conditions. It was reflected in the questionnaires filled in at the same time as the measurements of the physical parameters were taken. The following sections discuss this problem in detail.

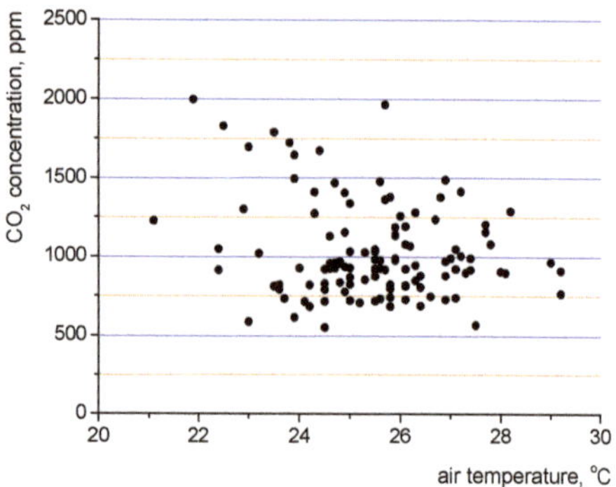

Figure 7. Indoor air temperature and carbon dioxide concentration in selected rooms between February 2017 and July 2018.

3.2. Thermal Sensation, Acceptability and Preference of the Occupants

The individuals expressed their responses to the thermal environment in the questionnaire survey. The average value for each room was calculated based on the answers. Thus, only rooms with at least five people were considered to generate the average (arithmetic mean) value of a given parameter. The first question in the questionnaire was: "How do you assess your current thermal sensation". The possible responses to choose from were: "too hot" (+3), "too warm" (+2), "pleasantly warm" (+1), "neutral" (0), "pleasantly cool" (−1), "too cool" (−2), "too cold" (−3). The comfort area ranged from −1 to +1. Other responses indicate unsatisfactory thermal comfort of the respondents. Figure 8 presents the combined results of 1369 responses as measured in 77 rooms (some were used more than once). Male and female votes were calculated separately, but placed in the same graph for comparison.

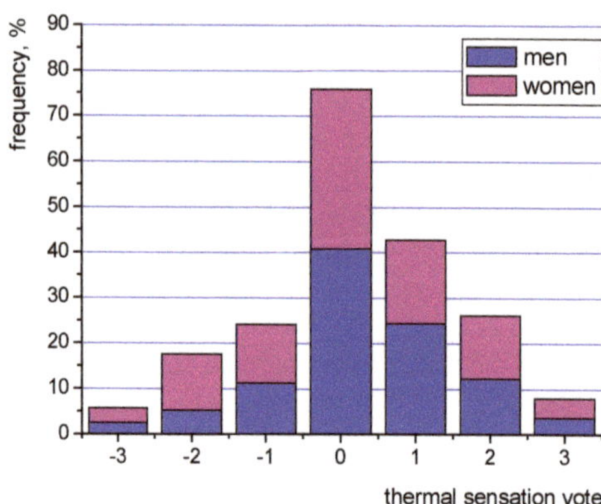

Figure 8. Thermal sensation vote of the room occupants in both intelligent buildings based on 1369 questionnaires (description in the text).

The research has shown that 66.3% of women and 76.4% of men felt comfortable (votes in the range of −1 to +1), while the rest felt uncomfortable (those defined as the total number of votes: +3, +2, −2, −3). The standards [6,8] stipulate that the percentage of people dissatisfied (PPD) with the microclimate should not exceed 10% for the buildings registered as category II. Thus, it can be concluded that generally the analysed modern intelligent buildings did not offer the proposed level of microclimate comfort conditions at the considered time span, during which the tests were performed. This phenomenon will be further considered in Figures 9 and 10, which present the thermal acceptability and thermal preference votes.

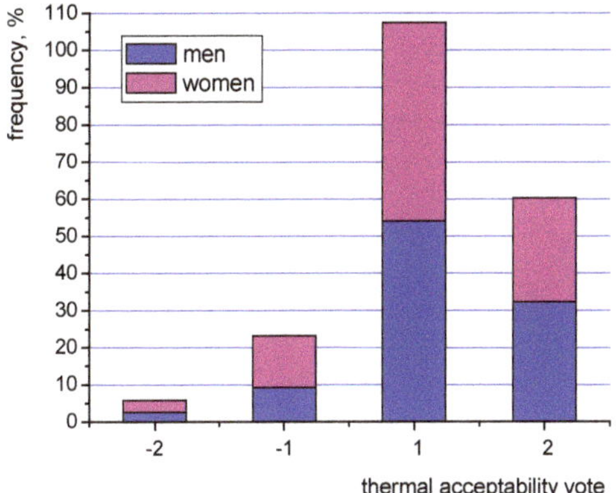

Figure 9. Thermal acceptability of the room occupants in both intelligent buildings based on 1369 questionnaires (description in the text).

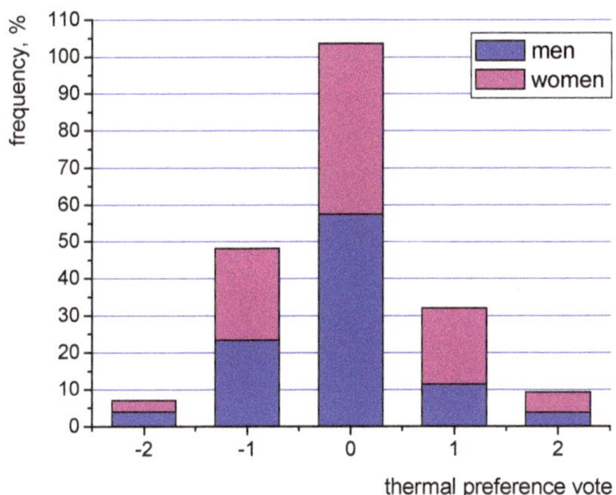

Figure 10. Thermal preference of the room occupants in both intelligent buildings based on 1369 questionnaires (description in the text).

The acceptability of room users was determined by them in the questionnaires as the answer to the question: "How do you rate your thermal environment?" The possible answers were: "absolutely acceptable" (+2), "still acceptable" (+1), "already unacceptable" (−1), "absolutely unacceptable" (−2). Figure 9 presents the combined results of 1369 responses.

Although Figure 8 suggested a large number of people dissatisfied, the acceptability vote (Figure 9) does not reflect that assumption completely. It has been determined that ca. 17% of women rated the environment as unacceptable (votes: −2 and −1) and only almost 12% of men. It has also been revealed that almost 28% of women and over 32% of men were absolutely pleased with the indoor environment conditions.

The questionnaire also contained the question about thermal preference. The occupants were asked to express their opinion on the following issue: "I would like in this room to be: "much warmer" (+2), "warmer" (+1), "no change" (0), "cooler" (−1), "much cooler" (−2). Figure 10 presents the combined results of 1369 responses as in the previous figures. The greatest number of respondents (ca. 46% of women and 57% of men) opted for "no change" answer. The others wanted a smaller or larger change. Many room users would like the environment to be cooler (ca. 24% for both men and women) or warmer (20.5% for women and 11.5% for men). The other votes, namely those of "much warmer" and "much cooler" did not exceed 5.5%.

The analysis of the data on thermal sensation, acceptability and preference indicate that the buildings, despite their sophistication and energy input, did not offer the highest level of thermal comfort. Since this parameter is considered to be mostly influenced by indoor thermal parameters, the following figures present the relationships between the operative temperature and thermal sensation (Figure 11), acceptability (Figure 12) and preference (Figure 13) in order to investigate the impact of operative temperature in detail. Here, average votes of thermal sensation, acceptability and preference have been considered for 77 groups of respondents in the considered rooms. Each group consisted of between 5 and 65 people.

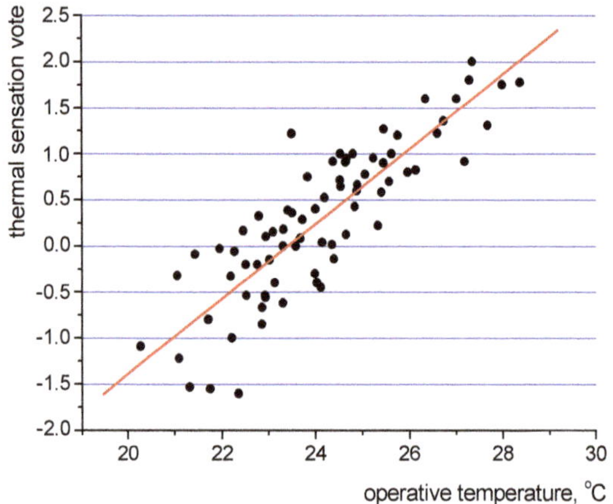

Figure 11. Relationship between average thermal sensation and operative temperature based on 1369 questionnaires.

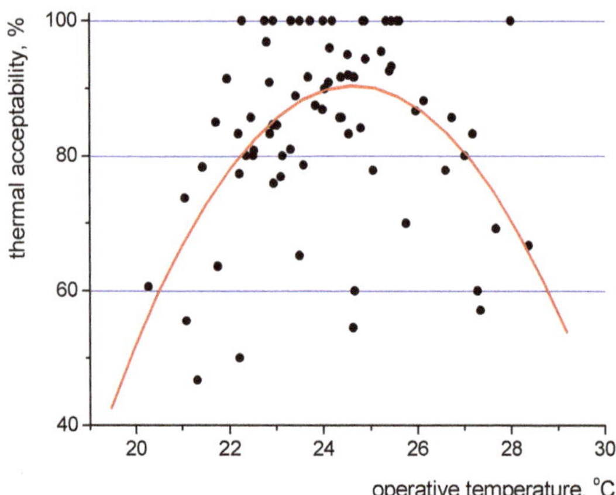

Figure 12. Relationship between average thermal acceptability and operative temperature based on 1369 questionnaires.

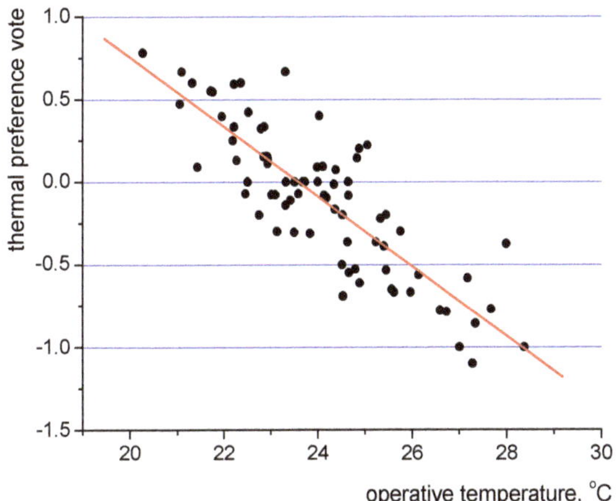

Figure 13. Relationship between average thermal preference and operative temperature based on 1369 questionnaires.

Based on the test results of a large number of people, it has been found that the relationship between the operative temperature and thermal sensation vote of the respondents is linear. The correlation takes the form of:

$$TSV = 0.4063T_{op} - 9.5123, \tag{1}$$

and is relatively strong for this type of research ($R^2 = 0.74$). It shows a significant influence of both air and radiant temperatures on thermal comfort. Other parameters, as will be discussed later, have a much lower impact. It is worth noting that Aghniaey et al. [25] has recently proposed a non-linear formula:

$$TSV = 0.055T_{op}^2 - 2.25T_{op} + 22.26, \tag{2}$$

However, the R^2 value was much lower (0.51).

The analysis of the acceptability of indoor conditions reveals that occupants voted overwhelmingly in favour of the thermal environment (over 80% of cases) when the operative temperature was in the range of 22–26.5 °C. The 90% level of acceptability was generally achieved for 22.5–25.5 °C (Figure 12), albeit with some exceptions.

The obtained fitting curve is characterised by the following formula:

$$TAV = -1.7782T_{op}^2 + 87.6605T_{op} - 989.92, \tag{3}$$

the research performed in [25] produced a correlation in the form of:

$$TAV = -0.06T_{op}^2 + 2.66T_{op} - 29.7, \tag{4}$$

however, only 15 measurement points were considered in [25].

The relationship between the thermal preference of the occupants and the operative temperature has been given in Figure 13. A strong linear correlation can be observed, indicating a crucial influence of air and radiant temperatures on this parameter.

The linear fitting procedure generated the following equation:

$$TPV = -0.2108T_{op} + 4.972, \tag{5}$$

whose R^2 was 0.71. In [25] a much lower value of R2 was obtained (0.52) for considerably fewer measurement data points, while the formula took the form of:

$$TPV = -0.05T_{op}^2 + 1.99T_{op} - 19.55. \tag{6}$$

A study on human thermal perceptions reveal other interesting facts. A relationship between thermal acceptability and thermal sensation votes has been presented in Figure 14. A strong correlation between these two parameters exists.

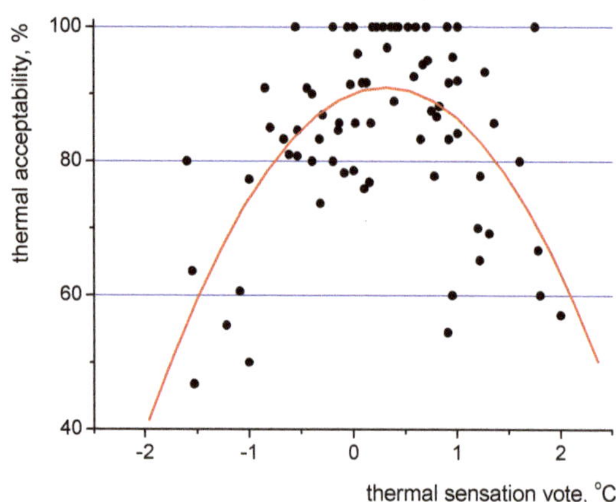

Figure 14. Relationship between average thermal acceptability and average thermal sensation vote based on 1369 questionnaires.

Based on the figure above, it can be concluded that the range of thermal sensation vote of −0.5–0.5, as proposed in the standards [6,8] for buildings category II, might not be the most preferable range. The large experimental data gathered in the project indicate that the respondents in the investigated intelligent buildings were in favour of a warmer environment. The most optimal range of thermal sensation vote proved to be −0.5–1.0.

It is also worth noting that people's preference vote—namely, the will to increase, decrease or maintain the temperature in the room—should be absolutely correlated with their thermal sensation vote. If the occupants consider the indoor environment to be "too hot" (+3) then their thermal preference should be: "much cooler" (−2). Similarly, if they rate the environment to be "too warm" (+2), their preference response should be: "cooler" (−1). If their thermal sensation is fine, they would require "no change" (0). The same situation would occur in the "cold" range of thermal sensation votes. Such a theoretical correlation has been presented in Figure 15 as the blue line. The figure also contains experimental data obtained during the study and their linear fit.

As can be seen in the figure both lines are in almost identical alignment, although the research was carried out in different buildings and throughout all the seasons. It proves that the testing procedure was performed correctly and that the proposed methodology of combining physical measurements and a questionnaire survey is very effective. Moreover, it can be concluded that the key element in thermal comfort analysis is the impact of air and radiant temperatures.

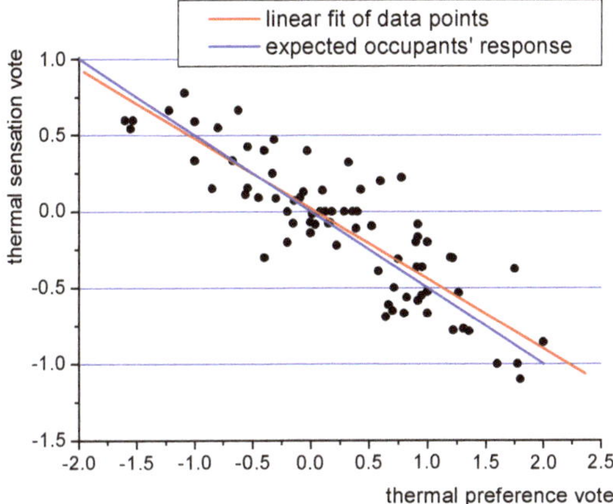

Figure 15. Relationship between average thermal sensation vote and average thermal preference vote based on 1369 questionnaires.

3.3. Thermal Comfort in the Lecture Theater

The largest lecture theater with 200 seats was selected for the analysis. At the time of the test 65 students were present there (39 women and 26 men). During the measurements, the indoor air temperature amounted to 26.8 °C, relative humidity 46.6% and the carbon dioxide concentration 1375 ppm. The occupants indicated their thermal sensation votes during the questionnaire survey. The mean value of TSV (thermal sensation vote) was 0.02. Thus, it can be concluded that the conditions in the room, despite quite a high air temperature, are optimal according to the standards [6,8], namely in the range of −0.5–0.5. Figure 16 presents the distribution of individual TSV values of the occupants.

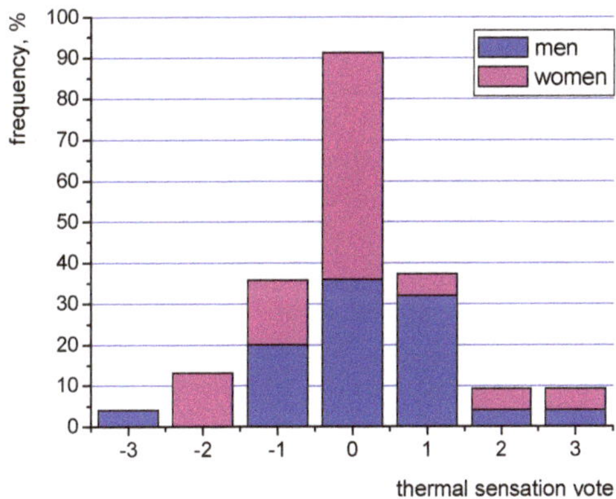

Figure 16. Thermal sensation votes in the lecture theater.

As can be seen the largest number of women (ca. 55%) and men (36%) rated the indoor conditions as very good (a neutral "0"). Generally, almost 85% of occupants expressed positive opinions (from "−1" do "+1") about their thermal sensations, which almost meets the requirements of the standards [6,8] of 90%.

Although, the mean thermal sensation was 0.02 the calculated results according to the standard [6] generated the PMV value of −0.09. The explanation of the discrepancy between the actual and predicted values might be linked with the influence of other parameters, not included in the methodology presented in the standard such as carbon dioxide concentration or BMI (body mass index) of individual respondents. Figure 17 presents the relationship between thermal sensation votes of the students in the lecture room 4.09 of the Energis building and their BMI values for the largest group of 65 people (1a) as well as two other groups, consisting of 60 (2a) and 57 people (3a).

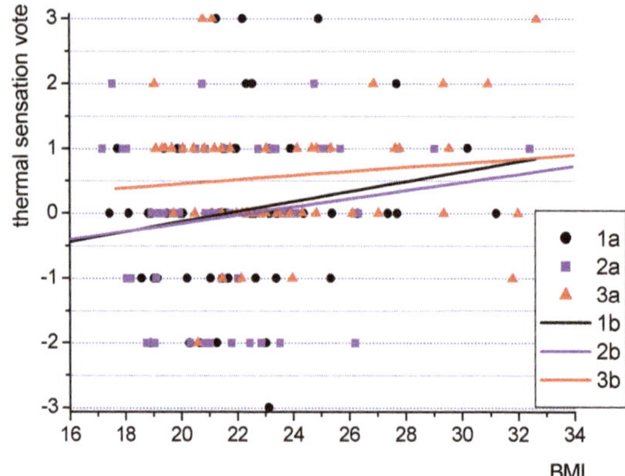

Figure 17. Relationship between thermal sensation votes and body mass index (BMI) for 3 different groups of people; 1a: survey on 65 people, 2a: survey on 60 people, 3a: survey on 57 people; 1b, 2b, 3b: linear fits of 1a, 2a and 3a, respectively.

There seems to be a weak dependence of BMI on thermal sensation of the respondents. People of large body mass index would generally consider the environmental conditions as warmer than those of low BMI. However, this issue needs further studies and clarification, especially considering that this problem is hardly ever considered in the literature.

3.4. General Sensation Vote

Apart from the thermal environment in the intelligent buildings, other factors such as noise, smell, and light intensity etc. might have an impact on the well-being of the occupants. It should also be considered as part of the study, since the analysed buildings ought to provide the overall positive sensations. Thus, a final question in the survey dealt with a general sensation of a respondent. Here, the volunteers were asked to provide the answer to the question on how they felt at the moment of filling in the questionnaire. The possible answers to choose from were: "very good" (+2), "good" (+1), "neutral" (0), "bad" (−1), "very bad" (−2). Figure 18 presents the distribution of the votes as taken from 1369 questionnaires.

The majority of both women and men were either satisfied with the environment ("+1") or had a natural feeling towards it ("0"). Some men and few women felt very good ("+2"). The rest, however, did not share those positive or neutral sensations. The negative feelings were expressed by ca. 18.5% of women and ca. 12% of men. Naturally, it might be related to various factors, other than thermal environment. In order to clarify the relationship between the general sensation and thermal sensation votes, a graph has been constructed (Figure 19) based on average data from 77 rooms to investigate the interdependence of these two parameters.

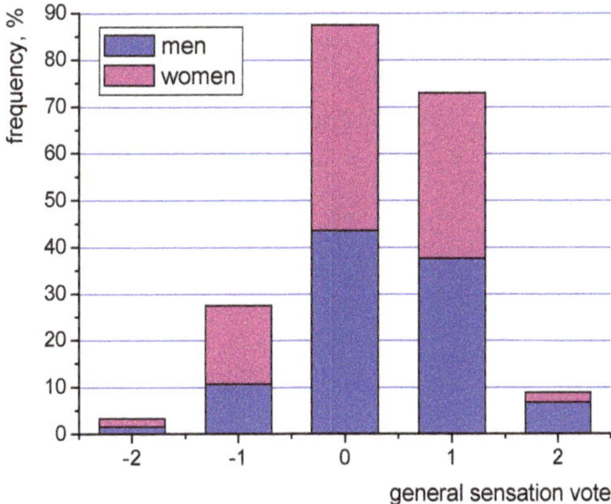

Figure 18. General sensation vote of the room occupants in both intelligent buildings based on 1369 questionnaires (description in the text).

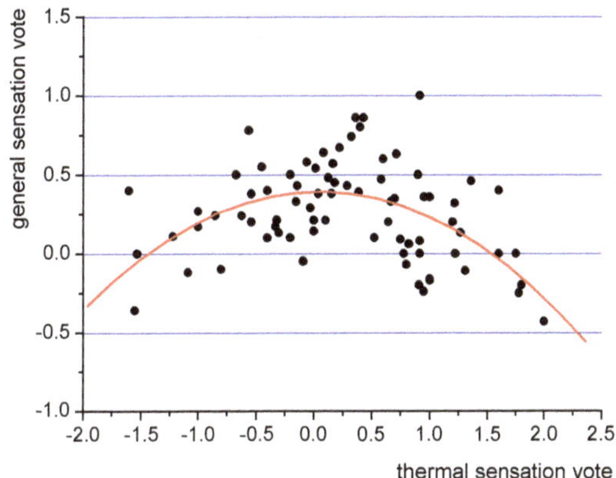

Figure 19. Average general sensation and thermal sensation votes based on 1369 questionnaires.

General sensation vote above 0 indicates the positive assessment of the environmental conditions. It occurs mostly in the range of thermal sensation votes from −0.75 to +1.25. It further proves the idea already mentioned in Section 3.2. that people tend to like warmer environments. At the same time, the respondents tend to assess the environment as negative (general sensation vote below 0), if the thermal conditions are perceived to be too hot or too cold, albeit with some exceptions. A deviation from this trend might be attributed to the influence of other factors, such as air humidity, carbon dioxide concentration as well as other individual factors such as BMI.

4. Conclusions

The tests performed in two modern intelligent buildings in the space of 1.5 years have generally shown that despite their sophistication and energy input, they did not offer the highest level of thermal comfort as evidenced by the data on thermal sensation, acceptability and preference of the respondents. General sensation vote, which is directly related to the quality of the indoor environment experienced by the occupants, has also proved to reveal that a number of people were unsatisfied.

The measurements of indoor air parameters showed that operative temperature sometimes exceeded the required value, although this might be caused by the occupants themselves and their thermal preferences. A similar situation occurred with relative humidity, whose value fell below 25%

during some measurements. Based on the thermal sensation vote it turned out that only 66.3% of women and 76.4% of men felt comfortable throughout the whole study. This is especially important, since thermal environment proved to be the most important element for ensuring the well-being of the occupants, as expressed by the general sensation vote of the occupants.

The obtained data have shown that people are generally in favour of a warmer environment. It might, to some extent, explain the number of people dissatisfied. Thus, the building operators might consider this fact for the settings of the HVAC systems. Actually, the role of the HVAC systems combined with BMS (building management system) is very important in ensuring high quality conditions. Their operation should be monitored constantly, so that comfort conditions are provided to the occupants at all times.

Broadening the experimental database will enable the development of control algorithms for the building management system that will be optimised to provide maximal thermal comfort conditions for building occupants while minimizing the energy consumption of modern and intelligent buildings. Such papers are rare and are usually based on computer simulations only [41], without backing the findings with real life research results. Thus, the future experimental work of the authors might have a scientific, but also a significant practical value.

Supplementary Materials: The following are available online at http://www.mdpi.com/1996-1073/13/8/1973/s1. Table S1: Data on room occupancy during the study in 77 rooms, Table S2: Data on carbon dioxide concentration measurements in 77 rooms, Table S3: Data on thermal sensation, thermal acceptability, thermal preference and general sensation votes in 77 rooms according to 1369 questionnaires.

Author Contributions: Conceptualization, Ł.J.O. and G.M.; methodology, Ł.J.O. and M.T.; software, Ł.J.O., N.R., J.P. and A.D.; literature review, Ł.J.O.; investigation, G.M.; resources, Ł.J.O., M.T. and N.R.; writing—original draft preparation, Ł.J.O.; writing—review and editing, Ł.J.O., G.M., M.T., funding acquisition, Ł.J.O., N.R., J.P. and A.D. All authors have read and agreed to the published version of the manuscript.

Funding: This research was funded by the funds from the Polish Ministry of Science and Higher Education.

Conflicts of Interest: The authors declare no conflict of interest.

References

1. Siano, P. Demand response and smart grids—A survey. *Renew. Sustain. Energy Rev.* **2014**, *30*, 461–478. [CrossRef]
2. Moujalled, B.; Cantin, R.; Guarracino, G. Comparison of thermal comfort algorithms in naturally ventilated office buildings. *Energy Build.* **2008**, *40*, 2215–2223. [CrossRef]
3. Merabtine, A.; Maalouf, C.; Waheed Hawila, A.A.; Martaj, N.; Polidori, G. Building energy audit, thermal comfort, and IAQ assessment of a school building: A case study. *Build. Environ.* **2018**, *145*, 62–76. [CrossRef]
4. Directive (EU) 2018/844 of the European Parliament and of The Council of 30 May 2018 Amending Directive 2010/31/EU on the Energy Performance of Buildings and DIRECTIVE 2012/27/EU on Energy Efficiency. *OJ* **2018**, *156*, 75–91.
5. Fanger, P.O. Calculation of thermal comfort: Introduction of a basic comfort equation. *ASHRAE Trans.* **1967**, *73*, 4.1–4.20.
6. ISO Standard 7730. *Ergonomics of the Thermal Environment—Analytical Determination and Interpretation of Thermal Comfort Using Calculation of the PMV and PPD Indices and Local Thermal Comfort Criteria*; International Organization for Standardization: Geneva, Switzerland, 2005.
7. *Thermal Environmental Conditions for Human Occupancy*; ANSI/ASHRAE Standard 55-2017; American Society of Heating, Refrigerating and Air-conditioning Engineering: Peachtree Corners, GA, USA, 2017.
8. CEN Standard EN 16798-1. *Energy Performance of Buildings—Ventilation for Buildings—Part 1: Indoor Environmental Input Parameters for Design and Assessment of Energy Performance of Buildings Addressing Indoor Air Quality, Thermal Environment, Lighting and Acoustics—Module M1-6*; European Committee for Standardisation: Brussels, Belgium, 2019.
9. Enescu, D. Models and indicators to assess thermal sensation under steady-state and transient conditions. *Energies* **2019**, *12*, 841. [CrossRef]

10. Šujanová, P.; Rychtáriková, M.; Sotto Mayor, T.; Hyder, A.A. Healthy, Energy-efficient and comfortable indoor environment, a review. *Energies* **2019**, *12*, 1414. [CrossRef]
11. Djamila, H.; Chu, C.M.; Kumaresan, S. Field study of thermal comfort in residential buildings in the equatorial hot-humid climate of Malaysia. *Build. Environ.* **2013**, *62*, 133–142. [CrossRef]
12. Guedes, M.C.; Matias, L.; Santos, C.P. Thermal comfort criteria and building design: Field work in Portugal. *Renew. Energy* **2009**, *34*, 2357–2361. [CrossRef]
13. Ricciardi, P.; Buratti, C. Thermal comfort in the Fraschini theatre (Pavia, Italy): Correlation between data from questionnaires, measurements, and mathematical model. *Energy Build.* **2015**, *99*, 243–252. [CrossRef]
14. Pala, U.; Oz, H.R. An investigation of thermal comfort inside a bus during heating period within a climatic chamber. *Appl. Ergon.* **2015**, *48*, 164–176. [CrossRef] [PubMed]
15. Yang, W.; Zhang, G. Thermal comfort in naturally ventilated and air-conditioned buildings in humid subtropical climate zone in China. *Int. J. Biometeorol.* **2008**, *52*, 385–398. [CrossRef] [PubMed]
16. Alamin, Y.I.; Del Mar Castilla, M.; Álvarez, J.D.; Ruano, A. An Economic model-based predictive control to manage the users' thermal comfort in a building. *Energies* **2017**, *10*, 321. [CrossRef]
17. Hong, S.H.; Lee, J.M.; Moon, J.W.; Lee, K.H. Thermal comfort, energy and cost impacts of PMV control considering individual metabolic rate variations in residential building. *Energies* **2018**, *11*, 1767. [CrossRef]
18. Robledo-Fava, R.; Hernández-Luna, M.C.; Fernández-de-Córdoba, P.; Michinel, H.; Zaragoza, S.; Castillo-Guzman, A.; Selvas-Aguilar, R. Analysis of the influence subjective human parameters in the calculation of thermal comfort and energy consumption of buildings. *Energies* **2019**, *12*, 1531. [CrossRef]
19. Almeida, R.M.S.F.; Ramos, N.M.M.; de Freitas, V.P. Thermal comfort models and pupils' perception in free-running school buildings of a mild climate country. *Energy Build.* **2016**, *111*, 64–75. [CrossRef]
20. Krawczyk, D.A.; Gładyszewska-Fiedoruk, K.; Rodero, A. The analysis of microclimate parameters in the classrooms located in different climate zones. *Appl. Therm. Eng.* **2017**, *113*, 1088–1096. [CrossRef]
21. Vilcekova, S.; Meciarova, L.; Burdova, E.K.; Katunska, J.; Kosicanova, D.; Doroudiani, S. Indoor environmental quality of classrooms and occupants' comfort in a special education school in Slovak Republic. *Build. Environ.* **2017**, *120*, 29–40. [CrossRef]
22. Fang, Z.; Zhang, S.; Cheng, Y.; Fong, A.M.L.; Oladokun, M.O.; Lin, Z.; Wua, H. Field study on adaptive thermal comfort in typical air conditioned classrooms. *Build. Environ.* **2018**, *133*, 73–82. [CrossRef]
23. Buratti, C.; Ricciardi, P. Adaptive analysis of thermal comfort in university classrooms: Correlation between experimental data and mathematical models. *Build. Environ.* **2009**, *44*, 674–687. [CrossRef]
24. Ricciardi, P.; Buratti, C. Environmental quality of university classrooms: Subjective and objective evaluation of the thermal, acoustic, and lighting comfort conditions. *Build. Environ.* **2018**, *127*, 23–36. [CrossRef]
25. Aghniaey, S.; Lawrence, T.W.; Sharpton, T.N.; Douglass, S.P.; Oliver, T.; Sutter, M. Thermal comfort evaluation in campus classrooms during room temperature adjustment corresponding to demand response. *Build. Environ.* **2019**, *148*, 488–497. [CrossRef]
26. Singh, M.K.; Kumar, S.; Ooka, R.; Rijal, H.B.; Gupta, G.; Kumar, A. Status of thermal comfort in naturally ventilated classrooms during the summer season in the composite climate of India. *Build. Environ.* **2018**, *128*, 287–304. [CrossRef]
27. Hens, H.S.L.C. Thermal comfort in office buildings: Two case studies commented. *Build. Environ.* **2009**, *44*, 1399–1408. [CrossRef]
28. Kuchen, E.; Fisch, M.N. Spot Monitoring: Thermal comfort evaluation in 25 office buildings in winter. *Build. Environ.* **2009**, *44*, 839–847. [CrossRef]
29. Ricciardi, P.; Buratti, C. Thermal comfort in open plan offices in northern Italy: An adaptive approach. *Build. Environ.* **2012**, *56*, 314–320. [CrossRef]
30. Indraganti, M.; Ooka, R.; Rijal, H.B. Thermal comfort in offices in summer: Findings from a field study under the 'setsuden' conditions in Tokyo, Japan. *Build. Environ.* **2013**, *61*, 114–132. [CrossRef]
31. Indraganti, M.; Ooka, R.; Rijal, H.B. Field investigation of comfort temperature in Indian office buildings: A case of Chennai and Hyderabad. *Build. Environ.* **2013**, *65*, 195–214. [CrossRef]
32. Jazizadeh, F.; Marin, F.M.; Becerik-Gerber, B. A thermal preference scale for personalized comfort profile identification via participatory sensing. *Build. Environ.* **2013**, *68*, 1440–1449. [CrossRef]
33. Szabo, J.; Kajtar, L. Thermal comfort analysis in office buildings with different air—Conditioning systems. *Int. Rev. App. Sc. Eng.* **2018**, *9*, 59–63. [CrossRef]

34. Fedorczak-Cisak, M.; Nowak, K.; Furtak, M. Analysis of the effect of using external venetian blinds on the thermal comfort of users of highly glazed offce rooms in a transition season of temperate climate—Case study. *Energies* **2020**, *13*, 81. [CrossRef]
35. Lipczynska, A.; Schiavon, S.; Graham, L.T. Thermal comfort and self-reported productivity in an office with ceiling fans in the tropics. *Build. Environ.* **2018**, *135*, 202–212. [CrossRef]
36. Zeiler, W.; Boxem, G. Effects of thermal activated building systems in schools on thermal comfort in winter. *Build Environ.* **2009**, *44*, 2308–2317. [CrossRef]
37. Cui, W.; Cao, G.; Ho Park, J.; Ouyang, Q.; Zhu, Y. Influence of indoor air temperature on human thermal comfort, motivation and performance. *Build. Environ.* **2013**, *68*, 114–122. [CrossRef]
38. Ghaffarianhoseini, A.; Berardi, U.; AlWaer, H.; Chang, S.; Halawa, E.; Ghaffarianhoseini, A.; Clements-Croome, D. What is an intelligent building? Analysis of recent interpretations from an international perspective. *Arch. Sci. Rev.* **2016**, *59*, 238–257. [CrossRef]
39. Leifer, D. Intelligent Buildings: A Definition. *Arch. Aust.* **1988**, *77*, 200–202.
40. Mishra, A.K.; Derks, M.T.H.; Kooi, L.; Loomans, M.G.L.C.; Kort, H.S.M. Analysing thermal comfort perception of students through the class hour, during heating season, in a university classroom. *Build. Environ.* **2017**, *125*, 464–474. [CrossRef]
41. Chen, X.; Wang, Q.; Srebric, J. Model predictive control for indoor thermal comfort and energy optimization using occupant feedback. *Energy Build.* **2015**, *102*, 357–369. [CrossRef]

Article

Multi-Criteria Optimisation of an Experimental Complex of Single-Family Nearly Zero-Energy Buildings

Małgorzata Fedorczak-Cisak [1,*], **Anna Kotowicz** [2], **Elżbieta Radziszewska-Zielina** [1], **Bartłomiej Sroka** [1], **Tadeusz Tatara** [1] **and Krzysztof Barnaś** [3,*]

[1] Faculty of Civil Engineering, Cracow University of Technology, 24 Warszawska Street, 31-150 Krakow, Poland; eradzisz@izwbit.pk.edu.pl (E.R.-Z.); bsroka@l3.pk.edu.pl (B.S.); ttatara@pk.edu.pl (T.T.)

[2] Colliers International, 18A Pawia Street, 31-154 Krakow, Poland; anna.kotowicz@colliers.com

[3] PhD Student, Faculty of Architecture, Cracow University of Technology, 24 Warszawska Street, 31-150 Krakow, Poland

[*] Correspondence: mfedorczak-cisak@pk.edu.pl (M.F.); krzysztof.k.barnas@gmail.com (K.B.); Tel.: +48-662-247-240 (K.B.)

Received: 3 February 2020; Accepted: 11 March 2020; Published: 25 March 2020

Abstract: The Directive 2010/31/EU on the energy performance of buildings has introduced the standard of "nearly zero-energy buildings" (NZEBs). European requirements place the obligation to reduce energy consumption on all European Union Member States, particularly in sectors with significant energy consumption indicators. Construction is one such sector, as it is responsible for around 40% of overall energy consumption. Apart from a building's mass and its material and installation solutions, its energy consumption is also affected by its placement relative to other buildings. A proper urban layout can also lead to a reduction in project development and occupancy costs. The goal of this article is to present a method of optimising single-family house complexes that takes elements such as direct construction costs, construction site organisation, urban layout and occupancy costs into consideration in the context of sustainability. Its authors have analysed different proposals of the placement of 40 NZEBs relative to each other and have carried out a multi-criteria analysis of the complex, determining optimal solutions that are compliant with the precepts of sustainability. The results indicated that the layout composed of semi-detached houses scored the highest among the proposed layouts under the parameter weights set by the developer. This layout also scored the highest when parameter weights were uniformly distributed during a test simulation.

Keywords: sustainable building; energy efficiency; NZEB; sustainable development

---

1. Introduction

Designing and building a complex of single-family houses, with a holistic focus on sustainable development, and construction and occupancy costs, is a problem that is very difficult to solve. It is a multi-aspect problem, in which decision variables include, among others: selecting construction materials, building services, the parameters of the shape of the buildings, their placement on the site, the sequence in which individual buildings are to be constructed and the completion deadlines for individual tasks. During the occupancy stage, one should add an analysis of the impact of the change in energy costs. When considering the entire life-cycle of the buildings, renovation and repair costs should also be included, in addition to any remodelling and demolition expenses. In recent years, the focus on the energy efficiency and pollution emission of buildings has grown exponentially and has had an immense impact on the real estate market and property values. The paper [1] presents an analysis of the profitability of investing in a selection of three systems based on renewable energy sources, using a semi-detached house as an example.

One of the components of designing a complex of single-family houses is establishing its urban layout. Multi-criteria methods are proposed by scholars who specialise in the optimisation of spatial planning, urban and architectural design, as well as construction project organisation and management increasingly often. In practice, general urban layout planning decisions are typically made as a result of many different approaches and factors, as well as under the influence of numerous stakeholders who may have different goals and hierarchies of values. Decision-making at this level should be supported by decision-making support methods during various stages and at different scales of design and planning. This should particularly apply to land use management and planning, which has been widely discussed in [2]. The authors of [3] have discussed the problem of uncertainty mapping using Monte Carlo analysis plugged into the Analytic Hierarchy Process (AHP) method, using the case of Howth, Dublin, in this context. However, this approach primarily focused on allowing decision-makers to gain insight into how much expert opinions can be trusted, specifically focusing on the AHP method, which can be considered a limitation.

The use of multi-criteria decision-making methods to support the planning of complexes of buildings of various sizes—ranging from entire districts to smaller compounds—is a field that can be considered to be fragmented in terms of research focus. However, such methods have been used to analyse various aspects of designing building complexes, ranging from those of placement within the wider urban fabric, the layout of buildings within complexes themselves, construction costs, the sequence of constructing individual buildings, as well as matters related to sustainability. Therefore, the authors are of the opinion that a broad characterisation of this use is justified. In the field of spatial planning, numerous authors have discussed the use of geographic information system (GIS) tools in combination with multi-criteria analysis methods. The authors of [4] analysed the problem of the optimal placement of housing estates within the structure of the city. In [5], a fuzzy version of the Decision Making Trial and Evaluation Laboratory (DEMATEL) method, coupled with Covariance Matrix Adaptation Evolutionary Strategy (CMA-ES) and Weighted Linear Combination (WLC) was used to model the suitability of sites for specific forms of use. The authors of [6], in turn, performed a comprehensive and cross-sectional review of the use of multi-criteria analysis methods in the field in question. In [7], the Ordered Weighted Averaging method (OWA) was used in conjunction with GIS tools to analyse the suitability of the territory of the Shavur Plain in Iran for various forms of use. It can therefore be concluded that GIS tools offer considerable potential for application in conjunction with multi-criteria analysis methods and that this field merits further study, particularly in light of the constant development of said tools. However, the utility of GIS tools appears to largely be confined to their role as information repositories and many of the proposals remain vague on whether or not they intend to automate data collection from these systems in any way. Multi-criteria methods have also been used to determine and investigate various indicators associated with sustainability and resilience.

The authors of [8] investigated potential use of multi-criteria methods in assessing sustainability, noting that new alternatives constantly become available, which can make method choice and weight setting difficult. In [9], the authors proposed a process of selecting multi-criteria analysis methods for renewable energy projects, thereby highlighting the immense variety and complexity of decision problems in the field and that there is no single best method that can be applied to them all. The authors of [10] used an expert system based on multi-criteria decision-making support to assess the impact of climate change, proposing a new method called MCDM-based expert system for adaptation analysis under changing climate, abbreviated as MAEAC. The authors applied it to a large territory and primarily focused on policy analysis and its impact on climate change. A model of selecting a method of assessing sustainable urban revitalisation strategies was proposed in [11], while the authors of [12] discussed the use of multi-criteria methods in environmental planning and management, similarly as in [13], where the objective was to arrive at a decision-making process associated with a sustainable approach to energy. The authors of [14] used the Discrete Linear Programming method (DLP) to analyse development projects associated with urban revitalisation. A comprehensive overview of the suitability of the use of multi-criteria methods to assess building-integrated green technologies, ultimately using the AHP method, was featured in [15]. The same method was also used by the authors of [16] to investigate construction project management in which sustainability was deemed a key factor.

Multi-criteria methods have also found use in studies of urban and architectural design. In [17], the AHP method was used to organise an architectural competition held in Italy. The authors of [18] used the Evaluation Based on Distance from Average Solution (EDAS) and Step-wise Weight Assessment Ratio Analysis (SWARA) methods to assess the shape of the floor plan of a newly-designed building, while in [19] a hybrid model was proposed—based on game theory, the AHP method, the Simple Additive Weighting method (SAW), Multiplicative Exponential Weighting, the Technique for Order of Preference by Similarity to Ideal Solution method (TOPSIS), EDAS, New Additive Ratio Assessment (ARAS), Full Multiplicative Form, as well as the rules of Laplace and Bayes. This approach is particularly noteworthy as it explores the simultaneous use of numerous (six) multi-criteria analysis methods in a single problem, thereby highlighting the importance of experts' opinions. The author of [20] studied the suitability of land for infill development using multi-criteria methods supported by GIS tools. In [21], the authors holistically investigated the problem of decision-making during the pre-design stage of an architectural design project. The impact of the shape of the building on its energy efficiency and the costs of its construction was analysed in [22], while [23] found use in architectural and urban design. The authors of [24] used the TODIM method (a Portuguese acronym for Interactive Multi-criteria Decision Making) to conduct a multi-criteria analysis of housing properties.

In summary, it can be stated that very few approaches that enhanced existing methods that had specifically been used to address problems in the field of spatial planning, urban and architectural design can be found, particularly as the range of problems said methods are applied to varies greatly between each case.

The works of Tawfik et al. [25], as well as Koenig etal. [26] merit particular attention, as they directly explore the problem of the multi-criteria selection of an urban layout or its determination. The proposals by Koenig et al. concerning the automated generation of pre-defined urban layouts for various plots currently occupied by informal settlements is also highly interesting [27–29].

1.1. The Scale of the Problem

According to the Eurostat Agency [30], in 2017, as much as 33.6% of the population of the European Union resided in detached single-family houses, while 24.0% of the population lived in semi-detached single-family houses. According to data provided by the Polish General Construction Inspector's Office [31], in 2018, a total of 98,915 building permit applications were filed for residential buildings, out of which 93,714 building permits were issued for single-family houses (almost 94.8%), which constituted a 3% increase from the previous year [31]. The results of statistical studies presented by BPD Europe BV [32] indicate that 12% of respondents from The Netherlands, 16% of respondents from France and 17% of respondents from Germany who had been searching for a new place of residence pointed to a single-family house as a preferred option in a situation in which they would have to move to a different dwelling. This proves that single-family buildings remain popular in many developed countries, while in countries such as Poland they continue to enjoy very high popularity. The design of single-family building complexes can therefore be considered an important field, both to planners who design urban layouts for individual projects, as well as designers who work on projects composed of many such buildings.

The necessity of considering parameters associated with energy efficiency is dictated by, among other things, the current energy policy vector of the European Union, expressed by such documents as Directive 2010/31/EU on the energy performance of buildings [33]. This EU document introduced the following definition of nearly-zero energy buildings (NZEB): nearly zero-energy building' means a building that has a very high energy performance, as determined in accordance with Annex I. The nearly zero or very low amount of energy required should be covered to a very significant extent by energy from renewable sources, including energy from renewable sources produced on-site or nearby". However, every European Union Member State has adopted its own NZEB parameters, which are adapted to local conditions.

The implementation of the Directive into Polish state law has introduced a new standard of near zero-energy buildings (NZEBs). The document that has been amended to include the standard, as stipulated in the Directive, is the Ordinance on the Technical Conditions that Must be Met by Buildings and their Placement [34]. In the Polish definition of NZEBs there are two criteria. The first criterion is ensuring proper building envelope thermal insulation. The second criterion for NZEB buildings in Poland applies to Primary Energy (PE), which cannot exceed a set value for different types of buildings. For single-family residential buildings without cooling installations, the maximum PE value is 70 kWh/(m^2/annum). The buildings analysed in the article meet both criteria, thus being compliant with NZEB standards as stipulated in Polish regulations.

Selecting the appropriate orientation of buildings, their shape, construction materials, building services and shading system solutions leads to lower energy demand and costs and affects the comfort of use of the buildings themselves, reducing overheating. This subject has been discussed in, among others, [35–42]. It can therefore be argued that the application of research findings in the field of energy efficiency and thermal comfort optimisation in the design of buildings and complexes that enjoy significant popularity on the real estate market should be incorporated into current design methodology. It can also be argued that by increasing the energy efficiency of buildings forming market-preferred layouts one can contribute to increasing the popularity of such solutions in areas where detached or semi-detached houses are the most popular.

1.2. Cost Estimation Methods

Calculating the overall costs of construction is a difficult task. Typically, the overall costs of construction are divided into indirect and direct costs. However, this is a major oversimplification. Situations that can significantly affect the expenditures of the contractor can take place over the course of working on a construction project. Some of these are difficult to predict (e.g., natural disasters, economic and social changes), while others can be prepared for during a project's planning stage. The predicted additional costs can include penalties for failing to meet planned deadlines or penalties for failing to maintain the continuity of construction crew operation.

Furthermore, every project is subjected to certain technological and organisational constraints, such as the time during which construction crews or buildings are available, or technological and organisational requirements concerning ongoing work performed by a given crew on a specific building. Numerous cost estimation or total construction project cost optimisation methods have been developed. They usually utilise metaheuristic methods, such as simulated annealing, hybrid algorithms [43] and genetic algorithms [44]. Mathematical programming methods are also used, such as the Decision model for planning material supply channels in construction [45]. In [46], an innovative approach was proposed, in the form of a stochastic decision network. It enables the integration of random and decision-based aspects of planned projects. This approach allows the decision-maker to obtain a solution that is optimal in terms of the expected time and cost. They can do so by defining their preferences as to the results of the planned project and setting their risk aversion accordingly. In the case of multi-criteria analysis of time and cost, the decision-maker can also define weight values for the expected project time and cost. The methods listed above do not incorporate all of the essential costs of carrying out construction projects, such as direct costs, indirect costs, penalties for failing to meet contractual deadlines, as well as the costs of discontinuities in the work of construction crews. They also do not take into account technological and organisational constraints. Furthermore, metaheuristic methods involve lengthy calculations when compared to linear programming.

1.3. Previously Used Methods

Multi-criteria decision-making analysis methods that have been applied in urban design and spatial planning have been predominantly focused on large-scale planning. As Afshari and Vatanparast [6] have indicated, such proposals utilised the Preference Ranking Organisational method for Enrichment Evaluation (PROMETHEE and PROMETHEE-2) methods in combination with Hasse Diagram Technique (HDT), the AHP method, Fuzzy AHP, Weighted Linear Combination (WLC), a combination of the AHP and TOPSIS methods, the ELECTRE-2 method (ELimination Et Choix Traduisant la REalité), the EXPROM-2 method (an extended version of the PROMETHEE method), as well as evolutionary algorithms. Guarini, Battisti and Chiovitti [23] also pointed to the possible use of Multi-attribute Utility Theory (MAUT) and Measuring Attractiveness by a Categorical Based Evaluation methods (MACBETH).

The selection of the multi-criteria analysis method depends on the specificity of the decision problem at hand. For instance, in [47] the author referred to studies in which they used the ELECTRE III and AHP methods to analyse the possibility of using an alternative version of the BIPOLAR method to solve the problem of selecting a construction company for partnering cooperation during a construction project. The study included a calculation example and verified the results of previous studies by applying an alternative BIPOLAR method version. Ultimately, its use produced the same end result as when using two other methods.

Multi-criteria methods that incorporate co-dependencies between decision-making criteria and alternatives have seen considerable development in recent years. In [48] the authors analysed a multi-criteria problem of selecting a new function for historical buildings in light of the proposed selection criteria. Apart from the economic aspect and cultural heritage, these criteria also included sustainability. The Weighted Influence Non-linear Gauge System method (WINGS), which was extended to incorporate the uncertainty of expert opinions and their aggregation, was applied to model the structure of the problem and analyse the dependencies between alternatives and criteria. The method had to be extended because of the specificity of the problem discussed in the paper. The majority of previously used methods did not analyse interdependencies between decision-making criteria despite such interdependencies actually existing, as proven by the authors. Similarly, the authors of [49] proposed a multi-criteria hybrid model, using the Decision Making Trial and Evaluation Laboratory method (DEMATEL) and the Analytic Network Process (ANP) to select a utility function for the purpose of adapting the building of 'Stara Polana', located in Zakopane. Both in this publication and in [50], a group of experts and specialists used a questionnaire to assess dependencies between factors that affect the criterion of societal benefits and the benefits associated with preserving cultural heritage. These were some of the parameters selected for analysis and rating in association with the investigated alternative forms of adaptive reuse of the Stara Polana heritage site in Zakopane. The remaining parameters, associated with, among others, acoustics, vibrations or thermal comfort, were studied using specialist technical instruments. These interdisciplinary studies became a basis for a multi-criteria analysis performed using a proposed hybrid model of selecting a form of use for a heritage site adaptation project.

1.4. The Significance of the Method Proposed in the Article

The innovative method proposed in this paper is an important contribution to the professional toolkit of designers and can significantly aid real estate developers in tailoring their offering. The method presented by the authors includes elements and criteria that are important during many stages of the design process: the conceptual urban design and site plan stage, the conceptual and technical architectural stage, as well as the structural system design and building services design stages. It also pertains to project cost assessment in the context of building single-family residential NZEBs, which are an important part of the housing sector. This approach can be considered a holistic perspective of the design task and the project as a whole, which is both necessary and desirable from the point of view of sustainable design. The necessity of this approach has been confirmed by, among others, Garau and Pavan [51] in a broader context.

The method developed by authors assumes a multi-criteria analysis that includes aspects of sustainability, as well as the costs of the construction and occupancy of a complex of single-family houses. Sustainability parameters, such as the energy demand of the buildings and greenhouse gas emissions (depending on the given building variants and their siting) have been determined through analyses using the EnergyPlus tool [52]. The energy analysis took the life-cycle stages of the buildings into consideration. The cost parameter was determined by including direct and indirect costs, the costs of construction (taking technological and organisational constraints into consideration), as well as the costs of building occupancy—the authors performed calculations for different building types and urban layout alternatives. Indirect and direct costs, as well as construction costs, were determined using a specifically-tailored optimisation model and the concept of priority scheduling. Occupancy costs were calculated using energy demand and the discount rate. Using previously selected parameters as a basis, the authors carried out a multi-criteria analysis using the Weighted Aggregated Sum Product Assessment method (WASPAS). This made it possible to determine the optimal solution for the problem in question.

2. Methods

2.1. Method Overview

The design optimisation task analysed by the authors is an actual design problem. The analysed complex of forty experimental buildings was, at the time of the writing of thepaper, planned to be constructed in the municipality of Mogilany near Krakow, Poland. The developer must select one of several urban layout alternatives, which is to be optimal in terms of cost and energy demand minimisation, in addition to being compliant with the precepts of sustainable development and real estate market preferences. The urban layouts under analysis can differ in terms of building siting, as well as the size or type of buildings (detached, semi-detached, terraced). Criteria values were determined for these layouts. The criteria adopted for the analysis included: overall construction cost, building energy demand, greenhouse gas emissions associated with building occupancy, building thermal comfort and an assessment of the urban layout by potential buyers. After setting parameter values for all alternatives, the decision-maker (developer) can set the weights of each parameter. After preparing the data in the listed manner, the authors conducted a multi-criteria optimisation using the WASPAS method. A block scheme of the proposed method will be presented in Figure 1. The method of setting the values of individual criteria will be presented in the following sections of the article.

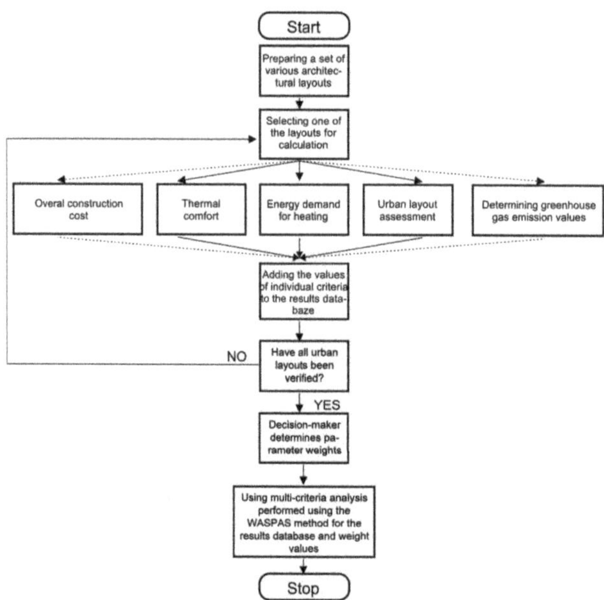

Figure 1. Method of selecting the urban layout of a complex of single-family houses.

2.2. Setting Parameters

2.2.1. Overall Construction Cost

Despite the contribution of the aforementioned methods, the approach prepared by the authors of this paper features numerous advantages. Its greatest advantage is that it describes all of the included cost dependencies (direct costs, indirect costs, penalties for failing to meet contractual deadlines, penalties for failing to ensure continuity in the work of construction crews) and constraints in the form of a linear programming model. This allows one to quickly find the optimal solution (e.g., using the Simplex method). If we allow the possibility of altering the sequence of building the structures, the rapid calculation of the goal function in the discrete optimisation algorithm can be considered a significant advantage. The proposed model permits the overlap of the work of construction crews on different plots, which is a good reflection of the working conditions of an actual construction site. Furthermore, the model allows for flexible adherence to technological and organisational constraints. This means that the model permits certain departures from the constraints defined by the designer. However, these departures are optimal (minimal). This is of particular significance in the case of projects with a large (even exceedingly large) number of constraints, as one can encounter a situation where it is impossible to meet all of them. The model's disadvantage is that it describes time-cost dependencies in linear form, which is a considerable simplification. Furthermore, the sequence of building the structures cannot be changed in the model. Both of these disadvantages will be studied further by the authors. The individual sections of the model have been discussed in [53,54].

The model has been defined as follows:

Given:

$$tgr_{i,j}, kgr_{i,j}, tn_{i,j}, kn_{i,j}, k_{indir}, kp_i, kc_j, Td_i, T_i^{SO}, T_i^{FO}, T_j^{SB}, T_j^{FB}, C_{i,j}^O, C_{i,j}^B \tag{1}$$

where:

$$C_{i,j}^O = \left(CT_{i,j}^{OL}, CW_{i,j}^{OL}, CT_{i,j}^{OU}, CW_{i,j}^{OU} \right) \tag{2}$$

$$C_{i,j}^B = \left(CT_{i,j}^{BL}, CW_{i,j}^{BL}, CT_{i,j}^{BU}, CW_{i,j}^{BU} \right) \tag{3}$$

with decision variables:

$$t_{i,j}, R_{i,j} Z_{i,j}, CK_{i,j}^{OL}, CK_{i,j}^{OU}, CK_{i,j}^{BL}, CK_{i,j}^{BU}, p_i, n_i \tag{4}$$

Goal function:

$$\overline{K} = K_{dir} + K_{indir} + K_p + K_c + K_{CO} + K_{CB} \rightarrow min \tag{5}$$

where:

$$K_{dir} = \sum_{i=1}^{n} \sum_{j=1}^{m} \left(t_{i,j} \cdot a_{i,j} + b_{i,j} \right) \tag{6}$$

$$K_{indir} = Z_{n,m} \cdot k_{indir} \tag{7}$$

$$K_p = \sum_{i=1}^{n} p_i \cdot kp_i \tag{8}$$

$$K_c = \sum_{j=1}^{m} \left(Z_{n,j} - R_{1,j} - \sum_{i=1}^{n} t_{i,j} \right) kc_j \tag{9}$$

$$K_{CO} = \sum_{i=1}^{n} \left(CW_{i,j}^{OL} \cdot CK_{i,j}^{OL} + CW_{i,j}^{OU} \cdot CK_{i,j}^{OU} \right) \tag{10}$$

$$K_{CB} = \sum_{j=1}^{m} \left(CW_{i,j}^{BL} \cdot CK_{i,j}^{BL} + CW_{i,j}^{BU} \cdot CK_{i,j}^{BU} \right) \tag{11}$$

Constraints:

$$t_{i,j} \geq tgr_{i,j} \tag{12}$$

$$t_{i,j} \leq tn_{i,j} \tag{13}$$

$$Z_{i,j} = R_{i,j} + t_{i,j} \tag{14}$$

$$R_{i+1,j} \geq Z_{i,j} + CT_{i,j}^{OL} - CK_{i,j}^{OL} \tag{15}$$

$$R_{i+1,j} \leq Z_{i,j} + CT_{i,j}^{OU} + CK_{i,j}^{OU} \tag{16}$$

$$R_{i,j+1} \geq Z_{i,j} + CT_{i,j}^{BL} - CK_{i,j}^{BL} \tag{17}$$

$$R_{i,j+1} \leq Z_{i,j} + CT_{i,j}^{BU} + CK_{i,j}^{BU} \tag{18}$$

$$Z_{i,m} - p_i \leq Td_i \tag{19}$$

$$R_{i,1} \geq T_i^{SO} \tag{20}$$

$$Z_{i,m} \leq T_i^{FO} \tag{21}$$

$$R_{1,j} \geq T_j^{SB} \tag{22}$$

$$Z_{n,j} \leq T_j^{FB} \tag{23}$$

$$t_{i,j}, d_{i,j,s}, R_{i,j}, Z_{i,j}, CK_{i,j}^{OL}, CK_{i,j}^{OU}, CK_{i,j}^{BL}, CK_{i,j}^{BU}, p_i \geq 0 \tag{24}$$

In order for the presented model to be applied, the following data is required(1)–(3):

- $tgr_{i,j}$–limit completion time for task performed on building i by crew j;
- $kgr_{i,j}$—limit costs of task performed on building i by crew j;
- $tn_{i,j}$—standard completion time for task performed on building i by crew j;
- $kn_{i,j}$—standard completion cost of task performed on building i by crew j;
- k_{indir}—indirect per unit cost of a multiple-structure project;
- kp_i—per unit cost of failing to meet a planned deadline for building i;
- kc_j—per unit cost of failing to maintain continuity of operation for crew j;
- Td_i—planned deadline for completing work on building i;
- T_i^{SO}—earliest availability window of building i;
- T_i^{FO}—latest availability window of building i;
- T_j^{SB}—earliest availability window of crew j;
- T_j^{FB}—latest availability window of crew j;

The parameters $C_{i,j}^{O} = \left(CT_{i,j}^{OL}, CW_{i,j}^{OL}, CT_{i,j}^{OU}, CW_{i,j}^{OU}\right)$ and $C_{i,j}^{B} = \left(CT_{i,j}^{BL}, CW_{i,j}^{BL}, CT_{i,j}^{BU}, CW_{i,j}^{BU}\right)$ are additional given data in the model, enabling the establishment of technological and organisational constraints. They have been described in the article at length [52].

The model's decision variables (4) are as follows:

- $t_{i,j}$—task duration (i,j);
- $R_{i,j}$—task initiation time (i,j);
- $Z_{i,j}$. —task completion time (i,j);
- p_i—delay value relative to the planned deadline for building i;
- in addition to variables that make it possible to establish technological and organisational constraints: $CK_{i,j}^{OL}$, $CK_{i,j}^{OU}$, $CK_{i,j}^{BL}$, $CK_{i,j}^{BU}$.

The goal function (5) is the sum of all essential project costs. Direct cost K_{dir} (6) is the sum of the costs of all works performed by all crews on all buildings, depending on the completion time of a given task. Direct cost includes the cost of labour, materials and equipment. It is assumed that a reduction in the task completion time, from a standard completion time to a limit time, is possible, but requires additional expenses. The dependency between the cost of completing a given task depends on the set standard and limit costs and is assumed to be linear. Therefore, the popular Critical Path Method Cost (CPM-COST) approach was incorporated into the model [55].The CPM-COST method enables the calculation of minimum direct costs, assuming the ability to shorten task completion time by incurring additional expenses when a fixed contractual deadline is imposed. Element (6) of the goal function and constraints (12)–(18) apply the assumptions of the CPM-COST method, extended to include flexible time couplings. Had the model included only (6) and (12)–(18), with parameters CT and CK removed, with the addition of the condition of meeting a contractual deadline, it would have been a classical CPM-COST model. The values $a_{i,j}$ and $b_{i,j}$ can be calculated using equations (25) and (26).

$$a_{i,j} = \frac{kgr_{i,j} - kn_{i,j}}{tgr_{i,j} - tn_{i,j}} \tag{25}$$

$$b_{i,j} = kgr_{i,j} - a_{i,j}tgr_{i,j} \tag{26}$$

Indirect cost K_{indir} (7) depends on the value of indirect per unit cost or the completion time of the entire multiple-structure project. Indirect per unit cost can be assessed using either a detailed or indicator-based method.

Costs associated with failing to meet planned deadlines K_p (8) are set as the sum of the products of the per unit cost of failing to meet a deadline and the length of the delay for all buildings. the per unit cost of failing to meet a planned deadline can vary between buildings. The penalties for such are defined in the contract between the developer and the buyer and typically range between 0.05 and 1% of the contract value for every day of delay. For instance, in [56] the authors define the cost of failing to meet contractual deadlines as ca. 0.1% of contract value per day of delay. Additional conditions depend on the country and its legal environment.

Costs associated with pauses in the operation of crews K_c (9) are calculated as the sum of the products of the per unit cost of pauses in crew operation and the length of the pause for all crews. The pause length is calculated on the basis of the earliest time the crew can complete its tasks on all buildings, the earliest time of initiation of work on the first building and the task completion times of the crew on all buildings.

In addition, the goal function includes the costs associated with failing to meet flexible couplings between tasks performed on successive buildings performed by the same crew K_{CO} (10) as well as flexible time couplings between tasks performed by successive crews on each building K_{CB} (11). The impact value of these couplings on the goal function for both types of couplings is calculated as the sum of the product of failing to meet the lower value of the flexible time coupling and the unit value for coupling failure, as well as the product of failing to meet the upper value of the flexible time coupling and the per unit value for coupling failure.

The following constraints have been included in the model. Constraints (12) and (13) constrain the task completion time to the limits between limit time and normal time. Constraint (14) enables the calculation of task completion times. Constraints (15)–(18) make it possible to determine the value of failing to meet appropriate flexible time couplings. Constraint (19) makes it possible to calculate the value of failing to meet the deadline of completing a given building. Constraints (20)–(23) implement the assumptions concerning the availability of every building and every crew. All decision variables used in the model are to take on non-negative values (24).

The goal function and all constraints are linear, which makes this a linear programming model. The Simplex method was used to solve the model [57].

The model described above is used to optimise a pre-set sequence of constructing buildings. In general, this sequence can be modifiable. Changing the building completion sequence affects the time and cost of completing the project. The authors plan to expand the model so that it will be able to optimise the sequence of constructing buildings. The proposed model was implemented using the Python programming language. The program can be obtained from the authors.

2.2.2. Energy Consumption

One of the criteria of assessing near-zero energy buildings (NZEBs) that are applicable in Poland is Primary Energy (PE), which cannot exceed the values stipulated in [34], and which differ depending on building type. For single-family residential buildings without cooling installations, the maximum PE value is 70 kWh/(m^2/annum). The Primary Energy (PE) values for the analysed buildings are lower than the maximum values allowed by Polish law. Primary Energy values predominantly refer to the building's sources of heating and electric power. Final Energy (FE) is a more precise indicator of material and building services solutions, as it results from a building's overall energy performance balance. This is why this indicator was incorporated into the multi-criteria analysis presented in the article.

The energy consumption simulation for all of the analysed alternatives was performed using DesignBuilder, a program enabling dynamic simulations. It is based on an hourly calculation method. The calculation engine of the program is based on EnergyPlus. DesignBuilder uses EnergyPlus format hourly weather data to define external conditions during simulations. These hourly weather data sets are data derived from hourly observations at the chosen location—in this case, Krakow, Poland.

2.2.3. CO$_2$ Emissions

A building consumes various natural resources, including water, materials and energy, and releases many pollutants and emissions during its life-cycle, i.e., from the raw material extraction to the final treatment of waste from the building's demolition [58]. The authors assume that emissions generated by construction material production, the construction of the buildings or their demolition, will be highly similar for both building variants (detached, semi-detached). This is why only occupancy-related CO$_2$ emissions were incorporated into the analysis, particularly as it is concordant with building energy performance calculations stipulated in Polish law [34].

The CO$_2$ emission in EnergyPlus is calculated from Fuel Totals using a locally published CO$_2$ emission factor for the specific region—in this case, Krakow. The consumed fuel (electricity for lighting and gas for DHW and Heating) is multiplied by the carbon emission factor. Based on emissions factors entered for a specific location, EnergyPlus calculates the mass or volume of CO$_2$ (carbon dioxide) [52] The amount of CO$_2$ produced depends on the carbon content of the fuel, and the amount of heat produced depends on its carbon and hydrogen content. Because natural gas, which is mostly methane (CH$_4$), has a high hydrogen content, combustion of natural gas produces less CO$_2$ for the same amount of heat produced from burning other fossil fuels [59].

2.2.4. Thermal Comfort

The methodology for determining thermal comfort is based on PN-EN ISO 7730 [60]. The thermal insulation of clothing was determined based on the PN-EN ISO 9920:2009 standard [61].

According to the ANSI/ASHRAE Standard 55-2010, Thermal comfort analyses were processed for indoor winter clothing (I_{cl} = 1 clo) for a typical winter week period, and summer clothing (I_{cl} = 0.5 clo) for a typical summer period. The I_{clo} value of 0.7 is assumed for transitional periods over the year. The insulation of clothing was determined as the value for the transitional season of clothing worn indoors I_{clo} = 0.7 (clo). Determining the PMV (Predicted Mean Vote) indicator requires the calculation of the physical parameters of the analysed space, while considering energy expenditure and the thermal insulation properties of the clothes of the persons working inside. The calculated PMV value is compared with a seven-point psychophysical scale of thermal sensation which has been developed by Fanger [62] (Table 1). A comfortable environment in terms of microclimate (thermally neutral) can be considered to be within the 0.5 < PMV < +0.5 range [63]. Thermal comfort sensations on the Fanger scale have been presented in Table 1.

Table 1. The seven-point PMV scale by Fanger.

+3 hot		−1 slightly cool
+2 warm	0 neutral (thermal comfort)	−2 cool
+1 slightly warm		−3 cold

Thermal comfort evaluation analysis was performed for two periods:

- A typical summer week—a week identified by the weather data translator software as being typical of summer.
- A typical winter week—a week identified by the weather data translator software as being typical of winter.

The authors assumed that these periods will be suitable for the purpose of this paper as they represent weather conditions that are repeatable and most likely to happen every year. The Polish thermal comfort standard [60] does not indicate specific periods of the year for measurements.

Design Builder uses three different mathematical models to calculate Thermal Comfort: the one by Fanger (the Fanger Comfort Model), the Pierce Foundation model (the Pierce Two-Node Model), and the model developed by researchers from Kansas State University (the KSU Two-Node Model). For the purpose of this paper, the Fanger model was chosen and implemented for thermal comfort analysis based on the equation from ISO 7730. All PMV values are taken from dynamic simulation outputs.

The results of the thermal comfort simulation analysis will be compared with in-situ measurements when similar conditions occur once all of the buildings have been completed. One of the hottest days of the year was selected for the comparative analysis, i.e., 12.07.2018. The simulation was performed for the living rooms of each building.

2.2.5. Urban Layout

The selection of the urban layout of a housing complex must take the preferences of future residents into consideration. In order to determine the optimal layout, a survey was conducted among the potential users of the housing complex's buildings. The respondents were presented with three urban layouts (Layout A, Layout B and Layout C) and asked to select the optimal urban layout in terms of housing comfort, their preferences and convenience. The survey was filled out by 31 middle-aged persons (30–50 years of age). The survey featured a three-point scale: a rating of 1; a rating of 2; a rating of 3. A rating 1 was given to the layout considered to be the most desirable by the respondent, a rating of 2 was given to the next-to-best layout and a rating of 3 was given to the least desirable of the layouts from the point of view the adopted assessment criteria.

2.3. MCDM Analysis—The WASPAS Method

The parameter sets selected by the authors (construction cost, energy demand, CO_2 emission, thermal comfort, urban layout assessment) are diverse and cannot be directly compared. This is why the authors decided to utilise a multi-criteria decision-making approach. There are numerous MCDM methods, such as AHP, ANP, ELECTRE, BIPOLAR, TOPSIS and many others [64,65]. For this specific problem, the authors selected the WASPAS method. It is a method that combines two other methods: the Weighted Sum Model (WSM) and the Weighted Product Model methods (WPM) [66–68]. The WASPAS method was selected because of its mathematical simplicity and capacity to provide highly precise results when compared to the WSM and WPM methods. The WASPAS method is currently widely accepted as an effective decision-making support tool [68]. It is continuously being developed (e.g., by integrating a fuzzy approach) [69,70].

The WASPAS method requires a decision matrix to be prepared: $X = \left[x_{i,j}\right]_{m \times n}$ where $x_{i,j}$ is the value of alternative i in relation to criterion j, m is the number of alternatives and n is the number of criteria. Afterwards, matrix X is to be standardised according to the following formulae:

$$\bar{x}_{i,j} = \frac{x_{i,j}}{\max_i x_{i,j}} \text{ for criteria which are stimulants}$$

$$\bar{x}_{i,j} = \frac{\min_i x_{i,j}}{x_{i,j}} \text{ for criteria which are inhibitors}$$

In order to calculate the importance of alternative i, the following formula is to be used:

$$Q_i = \lambda \sum_{j=1}^{n} \bar{x}_{i,j} w_j + (1 - \lambda) \prod_{j=1}^{n} \left(\bar{x}_{i,j}\right)^{w_j}$$

The optimal value of parameter λ can be calculated based on the formula below:

$$\lambda = \frac{\sigma^2\left(Q_i^{(2)}\right)}{\sigma^2\left(Q_i^{(1)}\right) + \sigma^2\left(Q_i^{(2)}\right)}$$

where the values $\sigma^2\left(Q_i^{(1)}\right)$ and $\sigma^2\left(Q_i^{(2)}\right)$ can be calculated from the following formulae:

$$\sigma^2\left(Q_i^{(1)}\right) = \sum_{j=1}^{n} w_j^2 \sigma^2\left(\bar{x}_{i,j}\right)$$

$$\sigma^2\left(Q_i^{(2)}\right) = \sum_{j=1}^{n} \left(\frac{\prod_{j=1}^{n}\left(\bar{x}_{i,j}\right)^{w_j} w_j}{\left(\bar{x}_{i,j}\right)^{w_j}\left(\bar{x}_{i,j}\right)^{(1-w_j)}}\right)^2 \sigma^2\left(\bar{x}_{i,j}\right)$$

The approximate variance value for the standardised criteria is calculated using the formula:

$$\sigma^2\left(\bar{x}_{i,j}\right) = \left(0.05\bar{x}_{i,j}\right)^2$$

The criteria adopted for the analysis using the WASPAS method have been presented in Figure 2. Of note is the weight selection process in the WASPAS method. The proposed method has been formulated with the developer in mind and it is they who make autonomous decisions concerning weight values, based on their own preferences. The method is a decision-making support tool for use by the developer and models their preferences.

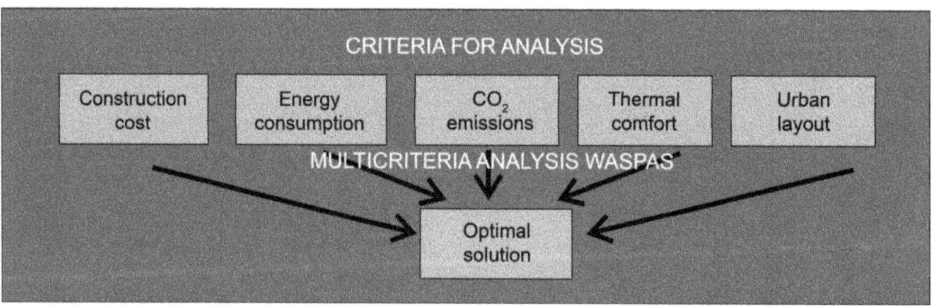

Figure 2. Criteria adopted for the WASPAS analysis.

In specific cases, the developer can use a different method of setting weights, e.g., using experts' opinions. The WASPAS method was applied by the authors using a Microsoft Excel spreadsheet. It is a simple solution that can be helpful to the developer. All that needs to be done is to set the weights and individual numerical values for parameters. The spreadsheet will calculate a score for each layout and produce a ranking of said layouts. The Excel file can be obtained from the authors.

3. Case Study—Housing Complex in Mogilany near Krakow, Poland

The planned housing complex which is the subject of the case study in this article is planned to be built on a 3-hectare plot in Libertów, in the municipality of Mogilany near Krakow, in the Lesser Poland Voivodship (Poland). The site is adjacent to an expressway which leads in the direction of Zakopane, a popular tourist destination located in the mountains. There is a lot of traffic along this road, particularly during weekends and holidays. The location of the experimental housing complex has been shown in Figure 3.

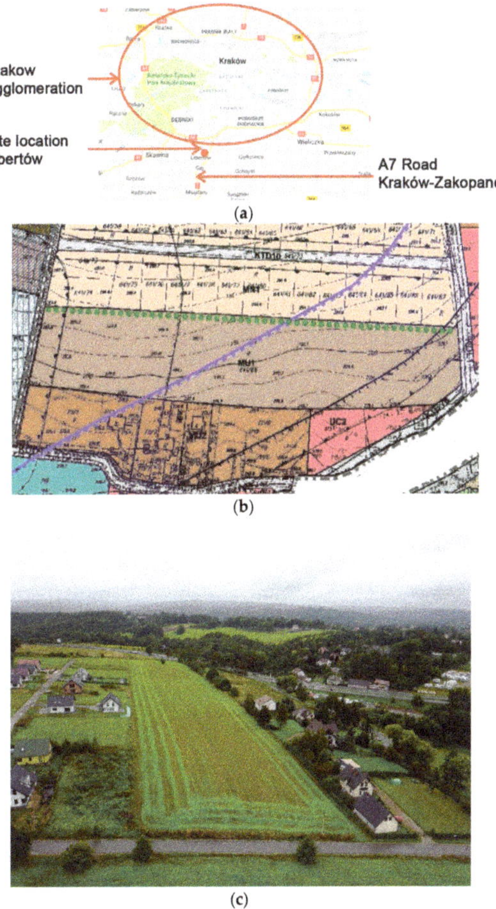

Figure 3. (**a**): Location of the experimental housing complex relative to the Krakow agglomeration; (**b**): Zoning plan of the plot assigned for the construction of the experimental housing complex; (**c**): View of the plot assigned for the construction of the experimental housing complex, as seen from a birds-eye view.

The site plan for the plot (MU1) has been shown in Figure 3b. Figure 3c shows an aerial view of the plot. The housing complex was assumed to be composed of 40 single-family residential buildings, with various building layouts (detached and/or semi-detached buildings).

Three proposals of the urban layout of the complex's buildings were analysed (Layout A, B and C). They will be presented in Figure 4. The expressway runs near the plot, to its south-east (SE). To the north of the plot is the direction towards Krakow, while from the south, to the holiday resort—Zakopane. The urban layouts that were analysed are a reflection of the preferences of persons who buy single-family houses in Poland in terms of their view of property desirability. Their selection is the result of the actual housing market situation in Poland. The proposed urban layouts are options that the developer offers for sale to their clients on the open market. The authors are aware of the precepts of sustainable design concerning the planning of small-scale urban complexes, but in this case the intent is to provide an iteration of the method for a more market-oriented approach. The authors do not rule out preparing a modification of the method that would feature other sustainability metrics in the future.

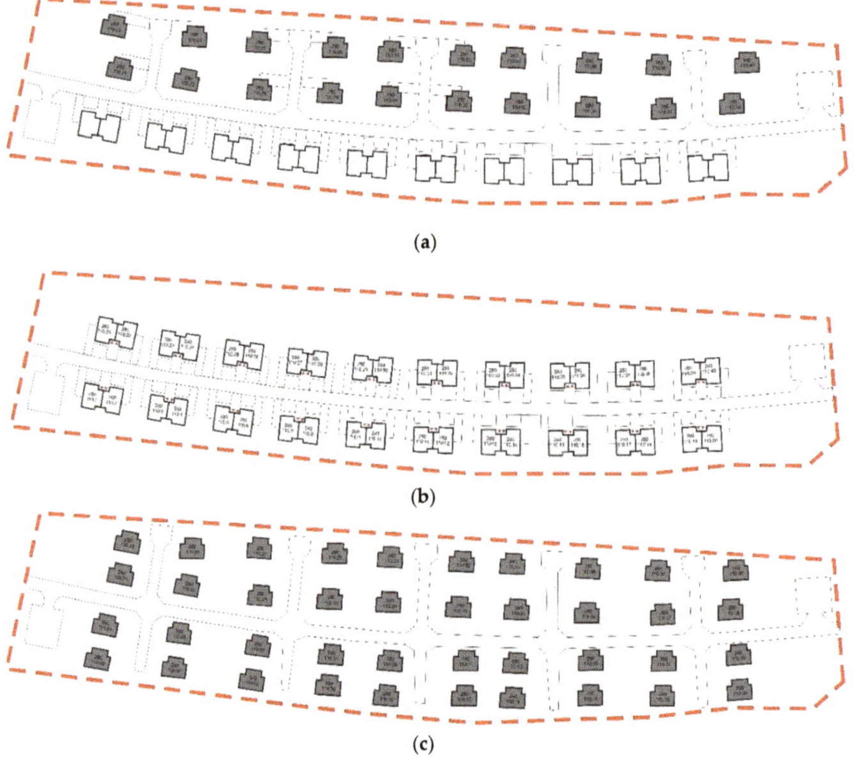

(a)

(b)

(c)

Figure 4. (**a**): Urban layout for alternative A; (**b**): Urban layout for alternative B; (**c**): Urban layout for alternative C.

Table 2 features a listing of the number of detached and semi-detached buildings for each layout alternative.

The housing complex will be composed of buildings with different floor areas and layouts. The types of individual buildings will be presented in Figure 5.

Table 2. List of buildings of each type for layout alternatives A, B and C.

Building Version/Description	Number of Buildings		
Layout alternative	A	B	C
W1/110m^2 detached	20	0	40
B1/110m^2 semi-detached	20	40	0

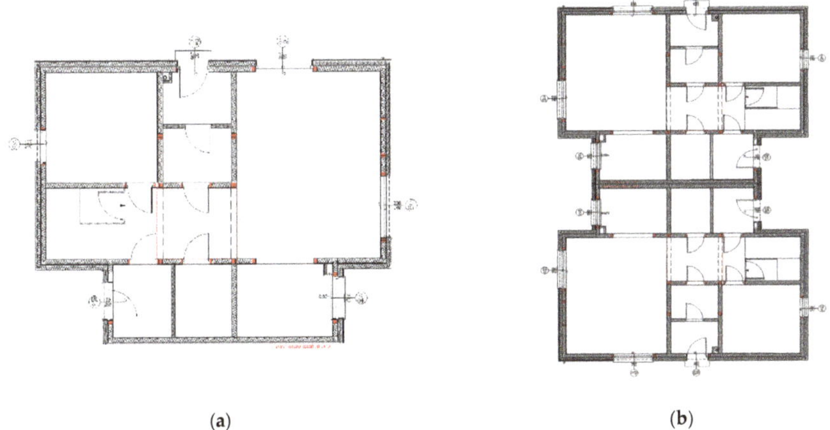

(a)   (b)

Figure 5. (**a**): Ground floor plan of a detached building with a usable floor area of 110 m^2 (variant W1); (**b**): semi-detached building (variant B1).

The buildings under analysis were assumed to be built using timber framing technology with thermal insulation composed of wood wool. The ground floor plan, assumed to have a usable floor area of 110 m^2, in a detached building, is to be presented in Figure 5a, and in a semi-detached building—in Figure 5b. The geometric models of the buildings subjected to the analysis have been presented in Table 3. B1 type buildings are linked with a common wall that separates the bathrooms of both buildings.

The cost of individual buildings was calculated based on "Biuletyn cen obiektów budowlanych (BCO) część I a—obiekty kubaturowe" (in English: Construction price bulletin, Part Ia—buildings) [71]. It is a cost bulletin that lists the mean prices of buildings in Poland. The BCO can be used to estimate the cost of constructing buildings or their fragments for the purposes of cost assessment or preparing financial schedules for construction projects. Direct cost was assessed using the construction price bulletin and includes labour, construction materials and machinery. The completion time for individual works was estimated on the basis of available material costs catalogues. The values for completion times and costs were assessed for both types of buildings featured in the case under analysis. Table 4 includes the values for a detached building with a usable floor area of 110 m^2.

Table 3. Geometric models of the analysed buildings.

Type	Usable Floor Area [m^2]	Geometric Model
W1	110	
B1	110 × 2	

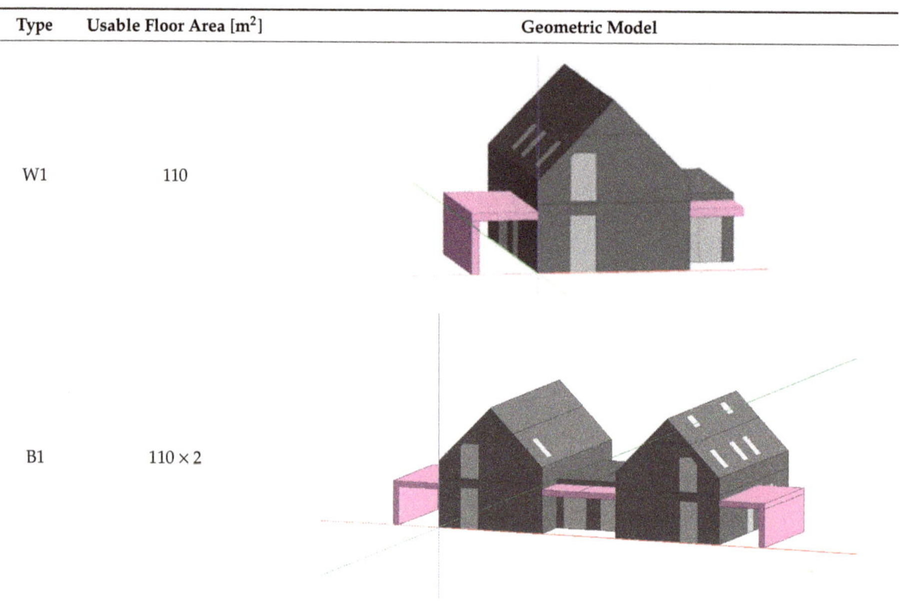

Table 4. Standard and limit costs of construction work and standard and limit construction work completion times, listed for individual works for a detached building with a floor area of 110 m^2.

Crew Number	Task Description	Standard Work Cost [Euro]	Total Limit Cost [Euro]	Standard Completion Time [Days]	Limit Time [Days]
1	2	3	4	5	6
1	Excavation	1100	1250	3	2
2	Footing, foundation wall and damp proof course	9150	10,150	12	8
3	Load-bearing walls, decks, roof and roofing	23,150	25,400	25	16
4	Building services assembly	16,100	17,400	17	11
5	Interior finishes (plasters, painting, door and window fitting, partition wall construction)	14,600	16,000	20	13
6	External finishes	3450	3850	10	6

In order to determine the overall construction costs, the following parameter values were assumed:

- indirect construction costs, estimated at 400 €/day; Indirect costs include the general costs of construction (wages for the site director, works managers, construction site infrastructure costs), as well as management costs (the cost of maintaining the project, company offices, costs of office work, etc.). Indirect costs were assessed as per the construction price bulletin. The bulletin lists the value of indirect costs at a level of ca. 15% of the total cost. The cost per day was calculated using an estimate of the project's duration.

- costs of failing to meet planned deadlines, estimated at 150 €/per day of delay; Penalties for failing to meet planned deadlines are defined in contracts between the developer and the buyer of the building. There is no official data on this subject and in the experience of the authors such penalties in Poland can range from 0.05% to even 1% of a contract's value per day of delay. It should be noted that penalties amounting to 0.5% of a contract's value per day are not considered unjustified by Polish courts of law. Furthermore, it should be noted that the penalty is not applied when the developer cannot be held responsible for the cause of the delay. It is impossible to factor in unforeseen events that can cause delays and for which the developer is not responsible during the construction planning stage, which is why this aspect was ignored in the paper. A penalty value of 0.2% per day per individual building was assumed in the paper.

- it was assumed that agreements with some contractors specified penalties for failing to ensure the continuity of crew operation. For the third crew, a 150 € penalty for each day of pause in the work was assumed, with 200 € per day assumed for the fourth crew. The penalty value for failing to ensure crew work continuity is a direct result of contracts with specific contractors. If a contractor performs a highly specialised section of the work, these penalties will be higher. This is why the third crew, which was assumed to assemble the innovative roof structure, and the fourth crew, which was assumed to handle building services installation, were assigned such high penalties for work discontinuity.
- it was determined that the first building is to be completed within 90 days. Every subsequent building was to be completed 14 days after the completion of the previous one. Planned deadlines are a direct consequence of contracts or agreements between potential buyers and the developer.

In addition, it was assumed that technological continuity was to be maintained between footing, foundation wall construction and insulation application, as well as the construction of load-bearing walls, decks, roof structure and roofing.

Table 4 includes the standard and limit costs of construction work, as well as the standard and limit completion times for individual works.

The optimisation process produced schedules for each of the analysed layouts that were optimal in terms of cost. For layout A, an overall cost of 3034 thousand euro was obtained, along with an overall completion time of 629 days. For layout B, the costs amounted to 2928 thousand euro, while the overall completion time was 629 days. For layout C, the costs amounted to 3233 thousand euro and the overall completion time was 680 days. The completion time was taken into consideration when calculating the costs of failing to meet planned deadlines for each building.

The buildings under analysis were designed to have a timber frame structure. The external partitions of the analysed buildings were assumed to be built using timber framing technology. The technology selection was primarily dictated by the developer's guidelines, who had the technical infrastructure enabling them to manufacture prefabricated timber elements. Figure 6 is to show a cross-section of the external wall.

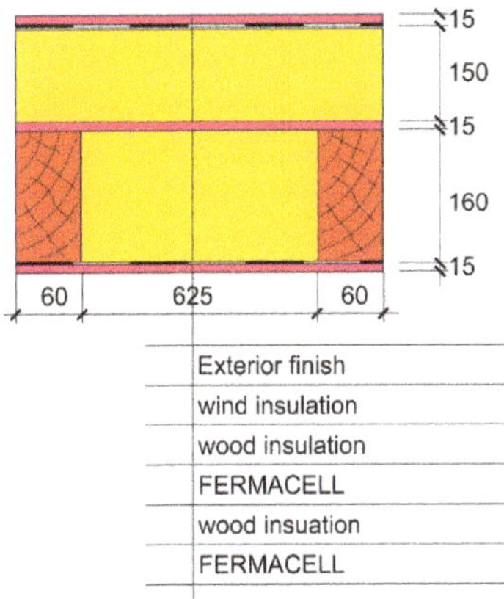

Figure 6. Cross-section through the external walls of the analysed buildings.

U heat transfer coefficients [W/(m^{2}K)] for elements of the buildings' envelope were calculated in accordance with the methodology outlined in [72,73]. The U heat transfer coefficients have been listed in Table 5, along with the standard requirements for NZEBs applicable in Poland [34].

Table 5. Heat transfer coefficients for elements of the buildings' envelope.

Item No.	Envelope Element	U Coefficient (Analysed Building) [W/(m^{2}K)]	U Coefficient (Requirements for NZEBs in Poland) [W/(m^{2}K)]
1	External wall	0.12	0.20
2	Roof	0.13	0.15
3	Slab on grade	0.10	0.30
4	External doors	1.3	1.3
6	Windows	0.8 (g = 0.64)	0.9
7	Roof windows	1.1 (g = 0.64)	1.1

The windows of the buildings meet Polish requirements for NZEBs. They were partially fitted with external blinds. The geometry of the buildings was also subjected to optimisation. Their massing is simple and thermal bridges were minimised. Their orientation relative to the cardinal directions is determined by local development guidelines. The structure of the buildings, material choices and their mass were determined by the developer. The air-tightness of the external envelope of the buildings was assumed to be n_{50} = 1.5 [1/h]. In Polish regulations concerning building design [34] there are two requirements concerning building air-tightness. For buildings with gravitational ventilation it is a value of 3.5 air exchanges per hour [1/h], while for buildings with mechanical ventilation, the number of air exchanges per hour is 1.5 [1/h]. This is why the authors assumed a value of 1.5 [1/h]. The analysis was performed during the buildings' design stage. At present, the first buildings have already been built. Air-tightness inspections performed on site indicate that the precise construction of the buildings has ensured a much greater air-tightness than initially assumed during the design stage. The air-tightness achieved during construction has a value of n_{50} = 0.6 [1/h], and is at a level expected of passive buildings [34]. The internal temperatures were assumed in accordance with the PN-EN 12831 standard [74].

Occupancy-related (metabolic) gains were defined in accordance with the ASHRAE 140/BESTEST standard [75], depending on the form of use of the spaces.

The buildings were assumed to be equipped with mechanical supply and exhaust ventilation with heat recovery. Supply air streams were assumed as in [76].

The Central Heating installation was designed in the form of gas-powered heating using a dual purpose gas-powered condensation boiler with a total efficiency rating of 0.95.

The indoor air temperature was assumed per PN-EN 12831:2006. Instalacje ogrzewcze w budynkach—Metoda obliczania projektowego obciążenia cieplnego—a Polish standard for heating installations governing methods of calculating heat loads during the design stage [74]. Per this standard, spaces for permanent occupancy by persons without outdoor clothing should be designed to have a temperature of +20 °C, while spaces meant for occupancy by persons potentially without clothing (such as bathrooms), should have a temperature of +24 °C. This temperature difference can be obtained by properly setting a building's thermostats or by using BMS installations. Concerning the schedule, it was assumed that such temperatures should be kept 24 h per day throughout the entire year. In residential buildings with building automatics systems it is much easier to assume periods of lowered temperature (e.g., at night) or when residents leave the building to work. However, most buildings in Poland do not feature elaborate BMS installations. The analyses assumed the buildings to have such systems, which is why no temperature drops were accounted for.

Design Builder Software uses Occupancy model data which defines the number of people in a space and occupancy times according to a set schedule. The Occupancy schedule setting is also used to control internal gains and/or HVAC systems. In DesignBuilder, schedules can be used to define:

- Occupancy times
- Equipment, lighting HVAC operation
- Heating and Cooling temperature setpoints

For the purpose of the simulation, the authors used a slightly modified Design Builder template with default data for residential buildings occupancy was used, to better reflect Polish living conditions, hours of work etc.

Lighting was designed in compliance with the PN-EN 12464-1:2012 standard [77]. Lighting was assumed to be energy-efficient, with a value of 3.3 [W/m^2] per 100 [lux]. The lighting intensity level was assumed as in [78].

The remaining parameters were assumed in accordance with the principles of construction, technical knowledge and applicable standards.

4. Results

4.1. Energy Consumption of Central Heating and Domestic Hot Water Preparation

The yearly final energy consumption for the purposes of heating, ventilation, domestic hot water preparation and lighting is to be presented in Figure 7. The yearly final energy consumption for the analysed buildings has been presented in Table 6.

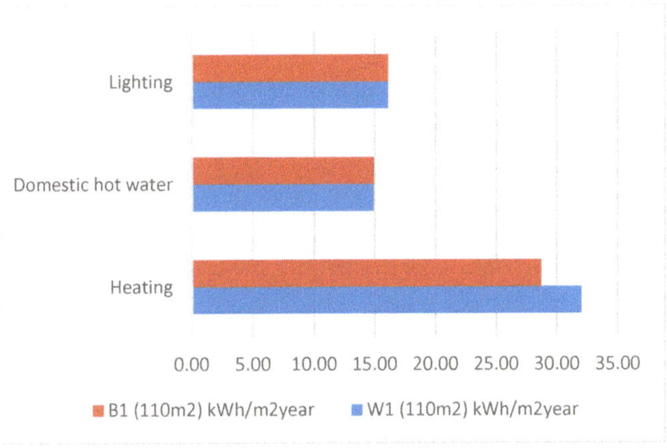

Figure 7. Yearly energy consumption for heating, domestic hot water preparation and lighting in the analysed buildings. B1 (semi-detached building), W1 (detached building).

Table 6. Yearly final energy consumption for the analysed buildings.

Building Type/Characteristic	Final Energy (FE) for Heating [KWh/(m^{2}annum)]	FE Domestic Hot Water [KWh/(m^{2}annum)]	Total FE [KWh/(m^{2}annum)]
W1/110 m^2 detached	32.0	14.9	46.9
B1/110 m^2 semi-detached	28.7	14.9	43.6

In the case of the 110 m^2 house (B1/110), its external wall touches its "twin", which results in an adiabatic zone and a heat transfer value of zero for these walls. This results in a lower final energy consumption for each pair of semi-detached buildings. Table 7 presents the yearly energy consumption for heating, ventilation and domestic hot water preparation for the entire complex.

Table 7. Summary for Layouts A, B and C, comprised of 40 buildings each.

Building Type/Characteristic	Layout A		
	Number of Buildings in the Layout	Energy Consumption for Heating [kWh/(m²annum)]	Total Energy Consumption [kWh/(m²annum)]
W1/110 m² detached	20	46.9	20 × 46.9 = 938.00
B1/110 m² semi-detached	20	43.6	20 × 43.6 = 872.00
TOTAL			1810.00
Layout B			
W1/110 m² detached	0	46.9	0.00
B1/110 m² semi-detached	40	43.6	40 × 43.6 = 1 744.00
TOTAL			1744.00
Layout C			
W1/110 m² detached	40	46.9	40 × 46.9 = 1 876.00
B1/110 m² semi-detached	0	43.6	0.00
TOTAL			1876.00

The high operational energy demand has its source in the Polish definition of NZEBs. The general definition of nearly-zero energy buildings is provided in the Energy performance of buildings directive 2010/31/UE as "'nearly zero-energy building' means a building that has a very high energy performance, as determined in accordance with Annex I. The nearly zero or very low amount of energy required should be covered to a very significant extent by energy from renewable sources, including energy from renewable sources produced on-site or nearby". However, every European Union Member State has adopted its own NZEB parameters, adapted to local conditions. In Poland, NZEBs are defined by the Ordinance on the Technical Conditions that Must be Met by Buildings and their Placement [34]. In the Polish definition of NZEBs there are two criteria. The first criterion is ensuring the proper building envelope thermal insulation. The heat transfer coefficients for external partitions in the analysed buildings are compliant with the values for NZEB buildings in Poland. The second criterion is for NZEBs in Poland pertains to Primary Energy value (PE), which cannot exceed a value set for different types of buildings. For single-family residential buildings without cooling, the maximum PE value is 70 kWh/(m²annum). In the article, the authors based their analyses on operational energy (OE), even though it is not listed in the requirements of the Polish NZEB definition. However, operational energy is more useful for the analyses performed by the authors than primary energy, which is a criterion for NZEBs in Poland. A definition of near-zero energy buildings for Poland has been included in the article. PE values were compared with maximum values for the Polish standard for NZEBs. The PE values for the analysed buildings are lower than the maximum values stipulated in Polish law, despite the fact that they are not fitted with renewable energy sources. This is why the authors define the buildings under analysis as NZEBs in their discussion. Table 8 shows the energy demand values for type A and type B buildings.

Table 8. Demand for different types of energy for type A and type B buildings

	Building A	Building B	Requirements for NZEBs in Poland
Building floor area [m^2]	114.2	114.2	-
FE_{H+W} [kWh/(m^2annum)]	46.9	43.6	-
FEpom [kWh/(m^2annum)]	5.9	5.9	-
wi (gas)	1.1	1.1	-
wi (electric power)	3.0	3.0	-
PE_{H+W} + pom [kWh/(m^2annum)]	$46.9 \times 1.1 + 5.9 \times 3.0 = 69.3$	$43.6 \times 1.1 + 5.9 \times 3.0 = 65.7$	70

4.2. CO_2 Emission in the Analysed Building Types

Calculation of CO_2 emissions was based on [52,59]. The calculation of CO_2 emissions is shown in Table 9.

Table 9. CO_2 emission calculations.

	Building W1	Building B1
Gas		
QK_{H+W} [kWh/annum]	5355.98	4979.12
QK_{H+W} [GJ/annum]	19.28	17.92
WE kg [CO_2/GJ]	55.41	55.41
$ECO_{2,\,H+W}$ [kg CO_2/annum]	1068.39	993.21
Electric Power		
QKpom [kWh/annum]	675.67	675.67
QKpom [GJ/annum]	2.43	2.43
WE [kg CO_2/GJ]	217	217
ECO_2, pom [kg CO_2/annum]	527.70	527.70
ECO2 [kg CO_2/annum]	1596.09	1520.91

The yearly CO_2 emission for the entire complex (40 buildings) has been presented in Table 10.

Table 10. Summary of CO_2 emissions per building type and for each layout alternative.

Building Type/Characteristic	CO_2 Emission per Building [kg/annum]	Number of Buildings Within the Complex	Total CO_2 Emission [kg/annum]
Layout A			
W1/110 m^2 detached	1596.09	20	31,921.80
B1/110 m^2 semi-detached	1520.91	20	30,418.20
	Total		62,349.00
Layout B			
W1/110 m^2 detached	1596.09	0	0
B1/110 m^2 semi-detached	1520.91	40	60,836.40
	Total		60,836.40
Layout C			
W1/110 m^2 detached	1596.09	40	63,843.60
B1/110 m^2 semi-detached	1520.91	0	0
	Total:		6843.60

CO_2 emissions are associated with the amount of fuel consumed for heating, as well as other types of fuel, such as electric power associated with, among other things, lighting. The low CO_2 emission of the buildings is also a product of their structural system. A reinforced concrete structure is associated with greater CO_2 emissions relative to a timber structure [78].

4.3. Thermal Comfort in the Analysed Building Layouts

The thermal comfort assessment analysis was carried out in a typical summer week, characteristic of climate data considered in this article (Southern Poland, around Krakow). The authors assumed that this period would be appropriate for the purposes of this paper, because it represents weather conditions that are repetitive and will most likely occur every year at a given location (this is also an option resulting from the software). The thermal comfort analysis includes building use schedules and HVAC system operation, which are discussed in the article. Graph 11 shows the daily average PMV values that are used for comparison between the buildings analyzed. Figure 8 presents the average values of thermal comfort for the analysed buildings.

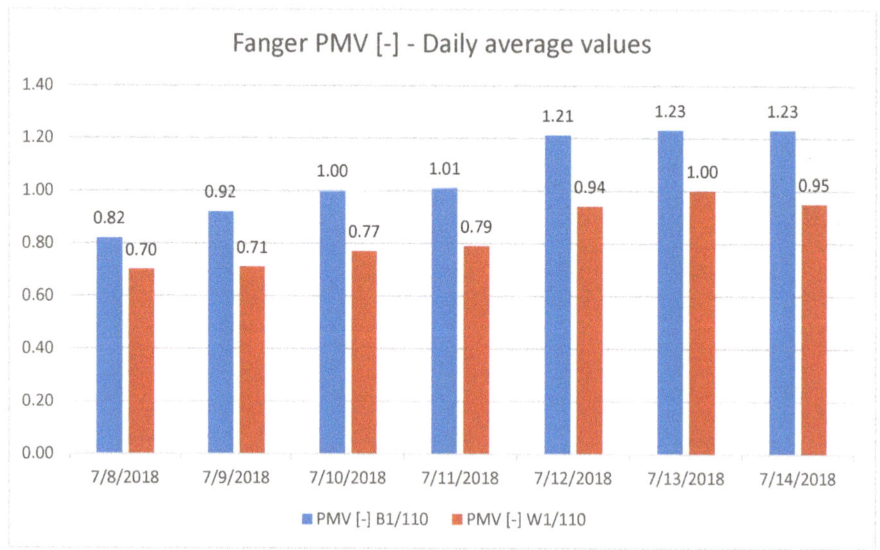

Figure 8. Thermal comfort for building B1/110.

The buildings were designed as low energy and have a high degree of air tightness. In addition, they are not equipped with external solar screens, so they receive high solar heat gains thanks to large glazed surfaces. The PMV value of 0.7—1.27 during a sunny and warm summer week can be considered a predictable result [79].

Overheating in buildings, as simulations show, is a real problem in heavily insulated, airtight buildings with windows facing south. It was assumed that no shading elements, such as blinds, venetian blinds or brise-soleils would be installed on buildings. Such assumptions were intentional and are part of another research project, which aims to determine the extent to which solar barriers improve the comfort of using buildings with almost zero energy consumption. The buildings analysed in the article did not have solar barriers, which is why their interiors were overheated.

In order to improve the thermal comfort of the analysed buildings, additional shading mechanisms should be added (external louvers, blinds, etc.). The shape of greenery around the buildings should be adjusted to increase the amount of air supplied to the mechanical ventilation or air conditioning installations. The authors calculated the average value of thermal comfort for each system to compare different urban layouts. The thermal comfort value for system A is 1.075, 1.21 for system B and 0.94 for system C.

4.4. Assessment of the Analysed Urban Layouts

The results of the survey performed among the potential residents of the experimental housing complex have been presented in Figure 9. The number of persons listing a given layout as the most desirable one is indicated by the blue column (no. 1), the red column shows the number of persons who rated the given layout as the second-most desirable (no. 2), while the number of persons who rated the layout as the least desirable has been shown in green (no. 3).

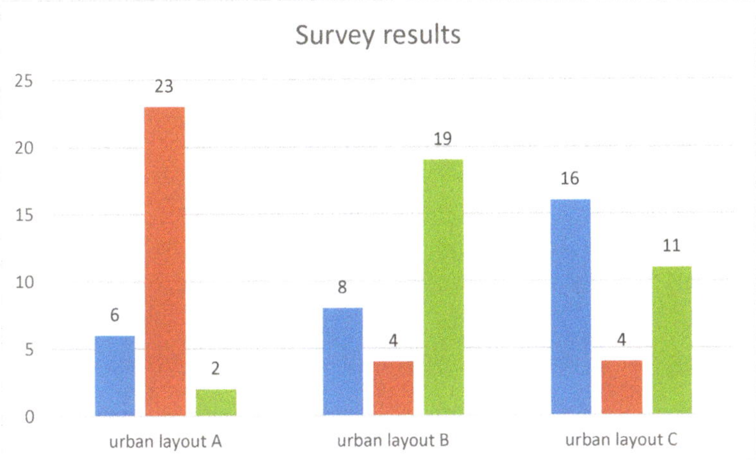

Figure 9. Survey results concerning the preferences of potential residents of the housing complex in Libertów.

The mean result of the surveys and the ranking of each layout has been presented in Table 11.

Table 11. Mean layout ratings as reported in the survey.

	Layout A	Layout B	Layout C
Mean value as reported in the survey	1.87	2.35	1.84
Rank	2	3	1

The respondents who selected Layout C as the most desirable reported that it was the layout that provided the most privacy (no adjacent neighbour). This layout was ranked as the most desirable by the greatest amount of respondents (16). Layout B was reported to be the least desirable by the largest amount of respondents (19). The respondents who rated this layout as the most desirable (rank 1) stressed their opinion as to lower construction and occupancy costs in semi-detached buildings when compared to other alternatives.

5. Results of the Multi-Criteria Analysis

Table 12 will present all of the results for each layout. As mentioned previously, the weights for each criterion, as included in the analysis, were set by the developer. They reflect the developer's preferences concerning each criterion. The developer decided that the most important criterion was total construction cost (a weight of 0.55). As the buildings will be sold to individual clients, their price was a key factor. The developer also took thermal comfort (a weight of 0.2) and energy consumption (0.15) into consideration, as this was intended to encourage potential clients and raise the price of buildings. The developer saw CO_2 emissions and urban layout assessment (each with a weight of 0.05) as the least important. Other weight values would produce different multi-criteria analysis results, which is why the authors also performed an analysis of the impact that weights can have on scores, as presented later in the paper.

Table 12. The numerical values entered into the WASPAS method for all layouts.

	Weight	0.55	0.2	0.15	0.05	0.05
		P1	P2	P3	P4	P5
		Construction Cost [Thousand Euro]	Thermal Comfort [-]	Overall Energy Consumption [kWh/(m²annum)]	CO₂ Emission [kg/annum]	Urban Layout Assessment [-]
	A	3034	1.075	1810	89,230.8	1.87
Alternative	B	2928	1.210	1744	87,750.0	2.35
	C	3223	0.940	1876	90,711.6	1.84
	Max	3223	1.21	1876	90,711.6	2.35
	Min	2928	0.94	1744	87,750	1.84

The scores for each urban layout produced by the analysis have been presented in Table 13. The scores have been presented on the scale of the entire complex (40 buildings). The authors used the WASPAS method to obtain the results. This made it possible to determine the optimal solution in light of the analysed factors (assessment criteria).

Table 13. Results of the multi-criteria analysis using the WASPAS method.

		Standardised Values											
		Construction Cost [thousand Euro]	Thermal Comfort [-]	Overall Energy Consumption [kWh/m²annum]	CO₂ Emission [kg/annum]	Urban Layout Assessment [-]	$\sum_{j=1}^{n} \bar{x}_{i,j} w_j$	$\prod_{j=1}^{n} (\bar{x}_{i,j})^{w_j}$	$\sigma^2(Q_i^{(1)})$	$\sigma^2(Q_i^{(2)})$	λ	Q_i	Rank
	A	0.965	0.874	0.964	0.983	0.796	0.939	0.938	0.000843	0.000604	0.418	0.938	2
Alternative	B	1.000	0.777	1.000	1.000	1.000	0.955	0.951	0.000885	0.000727	0.451	0.953	1
	C	0.908	1.000	0.930	0.967	0.783	0.927	0.925	0.000782	0.000607	0.437	0.926	3
	W_j	0.55	0.2	0.15	0.05	0.05							

We also performed a simulation for different criteria weight values using the WASPAS method. The randomly generated weights and their respective results are presented in Table 14.

Table 14. Simulation showing scores for different weight values.

						Score			Place		
lp	P1	P2	P3	P4	P5	A	B	C	A	B	C
1	0.303	0.258	0.248	0.139	0.052	0.934	0.940	0.938	3	1	2
2	0.325	0.062	0.158	0.234	0.220	0.925	0.985	0.902	w_j	1	3
3	0.098	0.069	0.273	0.284	0.276	0.915	0.984	0.901	2	1	3
4	0.160	0.175	0.239	0.187	0.239	0.910	0.959	0.909	2	1	3
5	0.203	0.380	0.146	0.089	0.184	0.900	0.912	0.927	3	2	1
6	0.103	0.370	0.140	0.210	0.177	0.904	0.914	0.934	3	2	1
7	0.295	0.041	0.116	0.342	0.205	0.931	0.990	0.908	2	1	3
8	0.043	0.253	0.361	0.130	0.213	0.906	0.941	0.918	3	1	2
9	0.127	0.259	0.277	0.151	0.187	0.911	0.939	0.922	3	1	2
10	0.163	0.245	0.236	0.223	0.133	0.923	0.943	0.931	3	1	2
11	0.383	0.155	0.269	0.161	0.031	0.948	0.963	0.933	2	1	3
12	0.203	0.335	0.264	0.062	0.137	0.911	0.922	0.930	3	2	1
13	0.076	0.233	0.260	0.279	0.153	0.921	0.945	0.931	3	1	2
14	0.211	0.019	0.348	0.341	0.081	0.955	0.996	0.927	2	1	3
15	0.265	0.334	0.164	0.209	0.028	0.933	0.922	0.951	2	3	1
16	0.281	0.266	0.169	0.183	0.101	0.926	0.938	0.933	3	1	2
17	0.234	0.216	0.130	0.189	0.231	0.908	0.949	0.911	3	1	2
18	0.342	0.105	0.320	0.075	0.158	0.928	0.975	0.908	2	1	3
19	0.197	0.146	0.059	0.224	0.374	0.891	0.966	0.887	2	1	3
20	0.194	0.201	0.178	0.215	0.213	0.913	0.953	0.915	3	1	2

Table 14. *Cont.*

						Score			Place		
lp	P1	P2	P3	P4	P5	A	B	C	A	B	C
21	0.458	0.048	0.310	0.048	0.137	0.937	0.989	0.904	2	1	3
22	0.234	0.211	0.214	0.308	0.032	0.945	0.950	0.946	3	1	2
23	0.315	0.073	0.273	0.080	0.260	0.914	0.983	0.892	2	1	3
24	0.142	0.279	0.131	0.259	0.188	0.911	0.935	0.927	3	1	2
25	0.320	0.197	0.022	0.353	0.108	0.934	0.954	0.933	2	1	3
26	0.296	0.263	0.021	0.233	0.188	0.912	0.939	0.921	3	1	2
27	0.321	0.274	0.137	0.239	0.030	0.939	0.936	0.946	2	3	1
28	0.109	0.295	0.091	0.214	0.291	0.891	0.931	0.911	3	1	2
29	0.176	0.035	0.145	0.242	0.401	0.896	0.992	0.877	2	1	3
30	0.149	0.127	0.295	0.299	0.130	0.935	0.970	0.926	2	1	3
31	0.317	0.143	0.097	0.124	0.320	0.898	0.966	0.889	2	1	3
32	0.356	0.037	0.362	0.110	0.135	0.939	0.991	0.908	2	1	3
33	0.228	0.216	0.152	0.289	0.114	0.930	0.949	0.933	3	1	2
34	0.305	0.019	0.151	0.193	0.332	0.908	0.995	0.882	2	1	3
35	0.164	0.089	0.311	0.116	0.321	0.903	0.979	0.888	2	1	3
36	0.060	0.163	0.311	0.124	0.343	0.892	0.962	0.892	2	1	3
37	0.231	0.185	0.250	0.282	0.052	0.943	0.956	0.940	2	1	3
38	0.091	0.197	0.091	0.242	0.379	0.886	0.954	0.893	3	1	2
39	0.112	0.290	0.257	0.158	0.182	0.909	0.932	0.925	3	1	2
40	0.129	0.063	0.259	0.297	0.252	0.920	0.985	0.904	2	1	3
41	0.048	0.056	0.152	0.377	0.368	0.902	0.987	0.891	2	1	3
42	0.080	0.418	0.045	0.368	0.090	0.918	0.903	0.957	2	3	1
43	0.187	0.204	0.191	0.213	0.204	0.914	0.952	0.916	3	1	2
44	0.223	0.311	0.353	0.046	0.067	0.925	0.928	0.938	3	2	1
45	0.204	0.325	0.141	0.262	0.068	0.928	0.925	0.947	2	3	1
46	0.047	0.331	0.208	0.081	0.335	0.878	0.923	0.903	3	1	2
47	0.459	0.331	0.110	0.033	0.066	0.923	0.923	0.934	2	3	1
48	0.385	0.117	0.029	0.205	0.264	0.912	0.972	0.897	2	1	3
49	0.176	0.314	0.096	0.226	0.188	0.907	0.927	0.927	3	2	1
50	0.144	0.293	0.063	0.072	0.428	0.866	0.932	0.885	3	1	2
						Sum of first ranks			0	40	10
						Sum of second ranks			27	5	18
						Sum of third ranks			23	5	22

6. Discussion and Conclusions

The results presented in Table 13 can aid the decision-maker in making decisions in selecting an architectural design alternative. The results were obtained for weights as defined by the developer and formulated to their individual preferences. An initial analysis of standardised values can provide valuable information about each alternative. Alternative B appears to clearly outrank the two other alternatives (highest value in four parameters). It also appears that alternative A is better than alternative C, as it has higher Q_i scores in all parameters except thermal comfort. The WASPAS method confirms this analysis. Alternative B was selected as the best, followed by alternative A, with alternative C ranked the lowest. The differences in Q_i scores were not significant (a difference of 1.5% and 2.8% relative to alternative B, for alternatives A and C, respectively).

An analysis of different weight values on the final results of the overall analysis was also performed (see Table 14). Generated with a uniform distribution, the weights and values from the Case Study were used to calculate Q_i score values in the WASPAS method. Statistically, alternative B was indicated to be the best (it ranked first in 80% of cases). Of note is that alternative C (ranked first in 20% of cases) was shown to be markedly better than alternative A (no first ranks). Therefore, it can be concluded that alternative A was dominated by alternatives B and C, which indicates that it can be excluded from the decision-making process. Alternative C outranked alternative B when a sufficiently high value (ca. 0.3) was assigned to the weight associated with thermal comfort. It is interesting that alternative

B received the lowest rank in 10% of cases. It was even outranked by alternative A when thermal comfort was assigned a weight with an appropriately high value (ca. 0.3) and urban layout assessment was assigned a low value (ca. 0.05). However, alternative C was shown to outrank the remaining alternatives each time when this was the case.

In the presented Case Study, only three alternatives of the design of the housing complex were formulated. The proposed model can be used to support decision-making problems with a greater number of alternatives. The method can also support the decision-maker in rejecting dominated alternatives, reducing the number of analysed alternatives.

It should be noted that analysing such complex cases, which incorporate parameters with different units, value ranges and significance, is impossible without using multi-criteria analysis methods. Despite the significant contribution of these methods to analysis, the decision-maker is irreplaceable. It is up to the individual to quantify their preferences and expectations and it is they who make the final decision. The method presented in the paper can only support them in this.

The work presents pioneering research associated with formulating a method of determining the optimal urban layout of a complex of single-family houses. The study took various cost factors into consideration. The method presented in the article is compliant with the concept of sustainable development. The article does not refer to the results of the work of other scholars as no direct counterparts were found in the literature. The discussion presented in the article covers the results of a multi-criteria assessment meant to aid in the selection of an optimal urban layout of timber single-family houses while taking different types of building layouts into account. The discussion concerning the results of the study took into consideration energy demand, the direct costs of construction and its organisation, occupancy costs, thermal comfort, energy consumption, greenhouse gas emissions and the results of a survey concerning preferences as to the proposed architectural and urban layouts conducted among potential home buyers. The authors assumed a goal function in the form of a sum of all essential costs, which, along with the constraints, is linear. The Simplex method was used to solve the linear model. Sustainability parameters were determined through analyses employing EnergyPlus code. Afterwards, using the procured data, a multi-criteria optimisation was performed, using the WASPAS method. The results of this optimisation made it possible to report a solution that can be considered optimal in light of the adopted assessment criteria.

Future Research Directions

Future analyses should automatically include the possibility of selecting construction materials, the shape of the buildings themselves and their placement. The user would define permissible alternatives and a computer program would, through discrete optimisation, set parameters that could prove useful in multi-criteria optimisation.

Author Contributions: Conceptualization, E.R.-Z. and M.F.-C.; methodology, E.R.-Z., B.S.; software, B.S., A.K.; validation, B.S., A.K., M.F.-C.; formal analysis, M.F.-C., B.S.; investigation, E.R.-Z., M.F.-C.; resources, M.F.-C.; data curation, T.T., E.R.-Z., M.F.-C.; writing—original draft preparation, B.S., A.K.; writing—review and editing, E.R.-Z., M.F.-C., K.B.; visualization, A.K., M.F.-C.; supervision, T.T., E.R.-Z., M.F.-C.; project administration and conclusion, T.T. All authors have read and agreed to the published version of the manuscript.

Funding: This research has received no external funding.

Conflicts of Interest: There are no conflicts of interest from authors.

References

1. Radziszewska-Zielina, E.; Rumin, R. Analysis of investment profitability in renewable energy sources as exemplified by a semi-detached house. In Proceedings of the International Conference on the Sustainable Energy and Environment Development, SEED 2016, E3S Web Of Conferences, Kraków, Poland, 17–19 May 2016; Volume 10, p. 00079.
2. Nijkamp, P.; Beinat, E. (Eds.) *Multi-Criteria Analysis for Land Use Management*; Kluwer Academic Publishers: Dordrecht, The Netherlands, 1998.
3. Quinn, B.; Schiel, K.; Caruso, G. Mapping uncertainty from multi-criteria analysis of land development suitability, the case of Howth, Dublin. *J. Maps* **2015**, *11*, 487–495. [CrossRef]
4. Blachowski, J.; Rybatkiewicz, W.; Warczewski, W.; Malczewski, P. Application of multi-criteria analysis in GIS for optimal planning of house development areas. Case study of Wrocław Functional Area. *Roczniki Geomatyki* **2016**, *14*, 561–571.
5. Zeydan, M.; Bostanci, B.; Oralhan, B. A New Hybrid Decision Making Approach for Housing Sustainability Mapping of an Urban Area. *Math. Probl. Eng.* **2018**. [CrossRef]
6. Afshari, A.R.; Vatanparast, M.; Cockalo, D. Application of multi-criteria decision making to urban planning—A review. *J. Eng. Manag. Comp.* **2016**, *6*, 46–53. [CrossRef]
7. Mokarram, M.; Aminzadeh, F. GIS-based Multicriteria Land Suitability Evaluation Using Ordered Weight Averages with Fuzzy Quantifier: A Case Study in Shavur Plain, Iran. *Int. Arch. Photogram. Remote Sens. Spat. Inf. Sci.* **2010**, *38 Pt II*, 508–512.
8. Cinelli, M.; Stuart, R.; Coles, K.K. Analysis of the potentials of multi criteria decision analysis methods to conduct sustainability assessment. *Ecol. Indic.* **2014**, *46*, 138–148. [CrossRef]
9. Kurka, T.; Blackwood, D. Selection of MCA methods to support decision making for renewable energy developments. *Renew. Sustain. Energy Rev.* **2013**, *27*, 225–233. [CrossRef]
10. Qin, X.S.; Huang, G.H.; Chakma, A.; Nie, X.H.; Lin, Q.G. A MCDM-based expert system for climate change impact assessment and adaption planning—A case study for the Georgia Basin, Canada. *Expert Syst. Appl.* **2008**, *34*, 2164–2179. [CrossRef]
11. Del Giudice, V.; de Paola, P.; Torrieri, F. An Integrated Choice Model for the Evaluation of Urban Sustainable Renewal Scenarios. *Adv. Mater. Res.* **2014**, *1030–1032*, 2399–2406. [CrossRef]
12. Lahdelma, R.; Salminen, P.; Hokkanen, J. Using Multi-criteria Methods in Environmental Planning and Management. *Environ. Manag.* **2000**, *26*, 595–605. [CrossRef]
13. Wang, J.-J.; Jing, Y.-Y.; Zhang, C.-F.; Zhao, J.-H. Review on multi-criteria decision analysis aid in sustainable energy decision-making. *Renew. Sustain. Energy Rev.* **2009**, *13*, 2263–2278. [CrossRef]
14. Nesticò, A.; Sica, F. The sustainability of urban renewal projects: A model for economic multi-criteria analysis. *J. Property Invest. Financ.* **2017**, *35*, 397–409. [CrossRef]
15. Si, J.; Marjanovic-Halburd, L.; Nasiri, F.; Bell, S. Assessment of building-integrated green technologies: A review and case study on applications of Multi-Criteria Decision Making (MCDM) method. *Sustain. Cities Soc.* **2016**, *27*, 106–115.
16. Erdogan, S.A.; Saparauskas, J.; Turskis, Z. A Multi-Criteria Decision-Making Model to Choose the Best Option for Sustainable Construction Management. *Sustainability* **2019**, *11*, 2239. [CrossRef]
17. Guarini, M.R.; D'Addabbo, N.; Morano, P.; Tajani, F. Multi-Criteria Analysis in Compound Decision Processes: The AHP and the Architectural Competition for the Chamber of Deputies in Rome (Italy). *Buildings* **2017**, *7*, 38. [CrossRef]
18. Juodagalviene, B.; Turskis, Z.; Saparauskas, J.; Endriukaityte, A. Integrated multi-criteria evaluation of house's plan shape based on the EDAS and SWARA methods. *Eng. Struct. Technol.* **2017**, *9*, 117–125. [CrossRef]
19. Turskis, Z.; Juodagalvienė, B. A novel hybrid multi-criteria decision-making model to assess a stairs shape for dwelling houses. *J. Civ. Eng. Manag.* **2016**, *22*, 1055–1065. [CrossRef]
20. Kamal, A. Suitability for Infill Development: A multi-criteria and Spatial Assessment Approach. In Proceedings of the EAAE-ARCC International Architectural Research Conference, Honolulu, HI, USA, 12–15 February 2014; pp. 336–345.
21. Elango, M.; Devadas, M.D. Multi-Criteria Analysis of the Design Decisions in Architectural Design Process during the Pre-Design Stage. *Int. J. Eng. Technol.* **2014**, *6*, 1033–1046.

22. Bostancioglu, E. Effect of building shape on a residential buildings construction cost, energy and life cycle costs. *Archit. Sci. Rev.* **2010**, *53*, 441–467. [CrossRef]

23. Guarini, M.R.; Battisti, F.; Chiovitti, A. A Methodology for the Selection of Multi-Criteria Decision Analysis Methods in Real Estate and Land Management Processes. *Sustainability* **2018**, *10*, 507. [CrossRef]

24. Uysal, F.; Tosun, O. Multi criteria analysis of the residential properties in Antalya using TODIM method. 2nd World Conference on Business, Economics and Management—WCBEM 2013. *Procedia—Soc. Behav. Sci.* **2014**, *109*, 322–326. [CrossRef]

25. Tawfik, H.; Nagar, A.; Fernando, T. Multi-Criteria Spatial Analysis of Building Layouts. *Issus Inf. Sci Inf. Technol.* **2007**, *4*, 781–789.

26. Koenig, R.; Standfest, M.; Schmitt, G. Evolutionary multi-criteria optimization for building layout planning. In Proceedings of the 32nd eCAADe Conference, Newcastle upon Tyne, UK, 10–12 September 2014; Volume 2, pp. 567–574.

27. Koenig, R.; Schmitt, G.; Standfest, M.; Chirkin, A.; Klein, B. Cognitive computing for urban planning. In *The Virtual and the Real in Planning and Urban Design*; Routledge: London, UK, 2017; pp. 93–111.

28. Koenig, R. Urban Design Synthesis for Building Layouts Based on Evolutionary Many-Criteria Optimization. *Int. J. Archit. Comput.* **2015**, *13*, 257–269.

29. Miao, Y.; Koenig, R.; Knecht, K.; Buš, P.; Chang, M.C. Computational urban design prototyping: Interactive planning synthesis methods—A case study in Cape Town. *Int. J. Archit. Comput.* **2018**, *16*, 212–226. [CrossRef]

30. Eurostat. Housing Statistics 2017. Available online: https://ec.europa.eu/eurostat/statistics-explained/index.php/Housing_statistics#Type_of_dwelling (accessed on 3 February 2020).

31. Główny Urząd Nadzoru Budowlanego, Ruch Budowlany w 2018 r. Available online: https://www.gunb.gov.pl/aktualnosc/ruch-budowlany-w-2018-r (accessed on 3 February 2020).

32. Joosten, H.; Wisman, H.; Klaver, S. Housing Markets in Perspective 2016, BPD Ontwikkeling BV 2016. Available online: https://www.bpdeurope.com/media/107467/q540_bpd_dunefra-2016_engels-lr-web.pdf (accessed on 3 February 2020).

33. Directive 2010/31/EU of The European Parliament and of The Council of 19 May 2010 on the energy performance of buildings. *Off. J. Eur. Union* **2010**, *18*, 13–35.

34. Rozporządzenie Ministra Infrastruktury w Sprawie Warunków Technicznych, Jakim Powinny Odpowiadać Budynki i ich Usytuowanie (Codified Text: Dz. U. RP 2019 pos. 1065 as amended). Available online: http://prawo.sejm.gov.pl/isap.nsf/download.xsp/WDU20150001422/O/D20151422.pdf (accessed on 3 February 2020).

35. Markiewicz, P. Smart Project & building—BIM-standard design of energy efficient buildings. *Hous. Environ.* **2017**, *21*, 72–82.

36. Fedorczak-Cisak, M.; Furtak, M.; Knap, K.; Kotowicz, A. Energy and cost analysis of windows in low-energy buildings: The influence of windows on the comfort of use of rooms. In Proceedings of the WMCAUS 2018: World Multidisciplinary Civil Engineering—Architecture—Urban Planning Symposium, Prague, Czech Republic, 18–22 June 2018.

37. Piasecki, M.; Kostyrko, K. Indoor environmental quality assessment, part 2: Model reliability analysis. *J. Build. Phys.* **2018**, *42*, 288–315. [CrossRef]

38. Zastawna-Rumin, A.; Nowak, K. Experimental thermal performance analysis of building components containing phase change material (PCM). *Procedia Eng.* **2015**, *108*, 428–435. [CrossRef]

39. Yarbrough, D.; Bomberg, M.; Romanska-Zapala, A. On the next generation of low energy buildings. *Adv. Build. Res.* **2019**. [CrossRef]

40. Romanska-Zapala, A.; Bomberg, M.; Yarbrough, D. Buildings with environmental quality management: Part 4: A path to the future NZEB. *J. Build. Phys.* **2019**, *43*, 3–21. [CrossRef]

41. Yarbrough, D.; Bomberg, M.; Romanska-Zapala, A. Buildings with environmental quality management, part 3: From log houses to environmental quality management zero-energy buildings. *J. Build. Phys.* **2019**, *42*. [CrossRef]

42. Romanska-Zapala, A.; Furtak, M.; Dechnik, M. Cooperation of Horizontal Ground Heat Exchanger with the Ventilation Unit During Summer—Case Study. In Proceedings of the World Multidisciplinary Civil Engineering-Architecture-Urban Planning Symposium—WMCAUS, IOP Conference Series-Materials Science and Engineering, World Multidisciplinary Civil Engineering-Architecture-Urban Planning Symposium (WMCAUS), Prague, Czech Republic, 12–16 June 2017. [CrossRef]

43. Rogalska, M.; Bożejko, W.; Hejducki, Z. Time/cost optimization using hybrid evolutionary algorithm in construction project scheduling. *Autom. Constr.* **2008**, *18*, 24–31. [CrossRef]

44. Agrama, F. Multi-objective genetic optimization for scheduling a multi-storey building. *Autom. Constr.* **2014**, *44*, 119–128. [CrossRef]

45. Ipsilandis, P.G. Multiobjective linear programming model for scheduling linear repetitive projects. *J. Cosntr. Eng. Manag.* **2007**, *133*, 417–424. [CrossRef]

46. Śladowski, G.; Szewczyk, B.; Sroka, B.; Radziszewska-Zielina, E. Using stochastic decision networks to assess costs and completion times of refurbishment work in construction. *Symmetry* **2019**, *11*, 398. [CrossRef]

47. Radziszewska-Zielina, E. The Application of Multi-Criteria Analysis in the Evaluation of Partnering Relations and the Selection of a Construction Company for the Purposes of Cooperation. *Arch. Civ. Eng.* **2016**, *2*, 167–182.

48. Radziszewska-Zielina, E.; Śladowski, G. Supporting the Selection of a Variant of the Adaptation of a Historical Building with the Use of Fuzzy Modelling and Structural Analysis. *J. Cult. Herit.* **2017**, *26*, 53–63. [CrossRef]

49. Fedorczak-Cisak, M.; Kowalska, A.; Radziszewska-Zielina, E.; Sladowski, G.; Pachla, F.; Tatara, T. A multicriteria approach for selecting the utility function of the historical building "Stara Polana" located in Zakopane. *MATEC Web Conf.* **2019**, *262*, 07002. [CrossRef]

50. Fedorczak-Cisak, M.; Kowalska-Koczwara, A.; Nering, K.; Pachla, F.; Radziszewska-Zielina, E.; Śladowski, G.; Tatara, T.; Ziarko, B. Evaluation of the Criteria for Selecting Proposed Variants of Utility Functions in the Adaptation of Historic Regional Architecture. *Sustainability* **2019**, *11*, 1094. [CrossRef]

51. Garau, C.; Pavan, V.M. Evaluating Urban Quality: Indicators and Assessment Tools for Smart Sustainable Cities. *Sustainability* **2018**, *10*, 575. [CrossRef]

52. *Big Ladder Software*; Engineering Reference—EnergyPlus 8.0,). Available online: https://bigladdersoftware.com/epx/docs/8-0/engineering-reference/ (accessed on 3 February 2020).

53. Radziszewska-Zielina, E.; Sroka, B. Priority Scheduling in the Planning of Multiple-Structure Construction Projects. *Arch. Civ. Eng.* **2017**, *64*, 21–23. [CrossRef]

54. Radziszewska-Zielina, E.; Sroka, B. Linearised CPM-COST model in the planning of construction projects. *Procedia Eng.* **2017**, *208*, 129–135. [CrossRef]

55. Jaśkowski, P.; Sobotka, A.; Czarnigowska, A. Decision model for planning material supply channels in construction. *Autom. Constr.* **2018**, *90*, 235–242. [CrossRef]

56. Sonmez, R.S.; Bettemir, Ö.H. A hybrid genetic algorithm for the discrete time-cost trade-off problem. *Exp. Syst. A* **2012**, *39*, 11428–11434. [CrossRef]

57. Nelder, J.A.; Mead, R. A Simplex Method for Function Minimization. *Comput. J.* **1965**, *7*, 308–313.

58. De Ruggiero, M.; Forestiero, G.; Manganelli, B.; Salvo, F. Buildings Energy Performance in a Market Comparison Approach. *Buildings* **2017**, *7*, 16. [CrossRef]

59. U.S. Energy Information Administration. Where Greenhouse Gases Come From. *Mon. Energy Rev.* **2017**. Environment. Available online: https://www.eia.gov/energyexplained/energy-and-the-environment/where-greenhouse-gases-come-from.php (accessed on 3 February 2020).

60. Polski Komitet Normalizacyjny. *PN-EN ISO 7730:2006 Ergonomia Środowiska Termicznego—Analityczne Wyznaczanie i Interpretacja Komfortu Termicznego z Zastosowaniem Obliczania Wskaźników PMV i PPD oraz Kryteriów Miejscowego Komfortu Termicznego*; PKN: Warszawa, Poland, 2006.

61. Polski Komitet Normalizacyjny. *PN-EN ISO 9920:2009 Ergonomia środowiska Termicznego—Szacowanie Izolacyjności Cieplnej i Oporu Pary Wodnej Zestawów Odzieży*; PKN: Warszawa, Poland, 2009.

62. Fanger, P.O. *Komfort Cieplny*; Arkady: Warszawa, Poland, 1974.

63. Klemm, P. *Budownictwo Ogólne—Fizyka Budowli*; Arkady Publishing House: Warszawa, Poland, 2010; pp. 54–55.

64. Suárez, R.; Escandón, R.; López-Pérez, R.; León-Rodríguez, A.; Klein, T.; Silvester, S. Impact of Climate Change: Environmental Assessment of Passive Solutions in a Single-Family Home in Southern Spain. *Sustainability* **2018**, *10*, 2914. [CrossRef]

65. Surahman, U.; Kubota, T.; Higashi, O. Life Cycle Assessment of Energy and CO_2 Emissions for Residential Buildings in Jakarta and Bandung, Indonesia. *Buildings* **2016**, *5*, 1131. [CrossRef]

66. Chakraborty, S.; Zavadskas, E.; Antucheviciene, J. Applications of WASPAS method as a multi-criteria decision-making tool. *Econ. Comput. Econ. Cybern. Stud. Res.* **2015**, *49*, 5–22.

67. Chakraborty, S.; Zavadskas, E.K. Applications of WASPAS Method in Manufacturing Decision Making. *Inform. Lith. Acad. Sci.* **2014**, *25*, 1–20.

68. Zavadskas, E.; Antucheviciene, J.; Šaparauskas, J.; Turskis, Z. MCDM methods WASPAS and MULTIMOORA: Verification of robustness of methods when assessing alternative solutions. *Econ. Comput. Econ. Cybern. Stud. Res.* **2013**, *47*, 5.

69. Turskis, Z.; Zavadskas, E.K.; Antucheviciene, J.; Kosareva, N. A hybrid model based on fuzzy AHP and fuzzy WASPAS for construction site selection. *Int. J. Comput. Commun. Control* **2015**, *10*, 873–888. [CrossRef]

70. Turskis, Z.; Goranin, N.; Nurusheva, A.; Boranbayev, S. A Fuzzy WASPAS-Based Approach to Determine Critical Information Infrastructures of EU Sustainable Development. *Sustainability* **2019**, *11*, 424. [CrossRef]

71. SEKOCENBUD. *Biuletyn cen Obiektów Budowlanych (BCO) część I a—Obiekty Kubaturowe*; SEKOCENBUD: Warszawa, Poland, 2019.

72. Polski Komitet Normalizacyjny. *PN-EN 6946 Komponenty budowlane i elementy budynku—Opór cieplny i współczynnik przenikania ciepła—Metoda obliczani*; PKN: Warszawa, Poland, 2017.

73. Polski Komitet Normalizacyjny. *PN-EN 13370 Cieplne właściwości użytkowe budynków—Przenoszenie ciepła przez grunt—Metody obliczania*; PKN: Warszawa, Poland, 2017.

74. Polski Komitet Normalizacyjny. *PN-EN 12831 Charakterystyka energetyczna budynków—Metoda obliczania projektowego obciążenia cieplnego—Część 1: Obciążenie cieplne*; PKN: Warszawa, Poland, 2017.

75. American Society of Heating, Refrigerating and Air-Conditioning Engineers. *Standard 140-2017—Standard Method of Test for the Evaluation of Building Energy Analysis Computer Programs (ANSI Approved)*; American Society of Heating, Refrigerating and Air-Conditioning Engineers: Atlanta, GA, USA, 2017.

76. Firląg, S. *Zrównoważone Budynki Biurowe*; PWN Scientific Publisher: Warszawa, Poland, 2018; Volume I, pp. 260–270.

77. Polski Komitet Normalizacyjny. *PN-EN 12464-1:2012 Światło i oświetlenie—Oświetlenie miejsc pracy—Część 1: Miejsca pracy we wnętrzach*; PKN: Warszawa, Poland, 2011.

78. Michałowski, B.; Marcinek, M.; Tomaszewska, J.; Czernik, S.; Piasecki, M.; Geryło, R.; Michalak, J. Influence of Rendering Type on the Environmental Characteristics of Expanded Polystyrene-Based External Thermal Insulation Composite System. *Buildings* **2020**, *10*, 47. [CrossRef]

79. Dudzińska, A.; Kotowicz, A. Features of Materials versus Thermal Comfort in a Passive Building. *Procedia Eng.* **2015**, *108*, 108–115. [CrossRef]

Article

Measurements of Energy Consumption and Environment Quality of High-Speed Railway Stations in China

Bin Qian, Tao Yu *, Haiquan Bi and Bo Lei

School of Mechanical Engineering, Southwest Jiaotong University, Chengdu 610031, China;
qianbin@my.swjtu.edu.cn (B.Q.); bhquan@163.com (H.B.); lbswjtu@163.com (B.L.)
* Correspondence: yutao073@swjtu.edu.cn; Tel.: +86-028-87600814

Received: 17 November 2019; Accepted: 23 December 2019; Published: 30 December 2019

Abstract: In recent years, the energy performance of public buildings has attracted substantial attention due to the significant energy-saving potential. As a semi-open high-space building, the high-speed railway station is obviously different from other public buildings and even traditional stations in terms of energy consumption and internal environment. This paper investigates the current energy consumption situation and environmental quality of 15 high-speed railway passenger stations in China. Results show that the energy consumption of the high-speed railway station is between 117–470 kWh/(m^2·a). The energy consumption of the station is related to the area and the passenger flow. The energy use of the station using district heating is higher than that of the station without district heating in the same region. The higher glazing ratio induces good natural lighting in the station, but the uniformity of the lighting in the station is not good. The acceptable temperature range of passengers in winter is larger than that in summer. The average air change rate of the high-speed railway station is 3.2 h^{-1} in winter and 1.8 h^{-1} in summer, which is the main reason of high energy consumption of the HVAC (Heating Ventilation Air Conditioning) system in this kind of building.

Keywords: high-speed railway station; energy consumption; lighting environment; thermal environment; air infiltration

1. Introduction

Buildings are responsible for approximately 40% of the total world annual energy consumption, most of which is used for the lighting and the HVAC (Heating Ventilation Air Conditioning) systems [1]. In China, building energy consumption has doubled in the past 20 years, with an annual growth rate of 3.7% [2]. Among them, energy consumption of public buildings accounts for 30% of energy consumed by all kinds of buildings [3]. Energy consumption of urban office buildings in China is about 33.6–107 kWh/(m^2·a), while the average energy consumption of commercial buildings is 55 kWh/(m^2·a) [4]. Due to the remarkable energy-saving potential, many studies [5–9] have focused on the energy-saving evaluation of public buildings. As an important and special branch of public buildings, traffic transportation buildings like railway stations have characteristics of large space span, various functions, high density of people, and long operation time of air conditioning system. Energy consumption and energy-saving approaches of these kind of buildings are also quite different from those of common public buildings [10].

Energy consumption of traditional railway stations in China is about 101–423 kWh/(m^2·a). Calculation results showed that the energy consumption of the HVAC system accounted for 30–60% of the total energy consumption of the station, while the lighting system and the elevator system had a percentage of 10–20% and 5–10% [10,11], respectively. Li et al. [12] studied the energy-saving potential of a single railway station by simulations. Results showed that the energy consumption of the waiting lounge of the railway station can be saved by 17–30% by changing the design temperature or the operation mode of HVAC systems. Because of the large amount of unorganized ventilation caused by the opening of the station building, the installation of the air curtain to organize the infiltration air can reduce 12% of the cooling load [13]. Relevant research on the elevator suggested that 30% of energy consumed by the elevator system can be saved in theory for typical stations through reasonable operations of the elevator system [14]. For the lighting system, as a kind of low power equipment, LED (Light Emitting Diode) light is widely used for lighting. Energy efficiency can be improved by controlling the lighting intensity reasonably according to the intensity of natural light as well as the arrival and the departure of vehicles [15–18].

Indoor thermal environment research has mainly focused on office buildings [19]. Tests showed that the acceptable comfort temperature in the building was a temperature range rather than a single temperature value [9]. A wider control temperature range can save energy, and it will not significantly decrease the indoor thermal comfort [20]. Expanding the temperature range of the thermostat can save the cooling system energy consumption by 31% and the heating energy consumption by 12% [21]. By using natural ventilation in different cities, the cooling energy consumption can be reduced by 8–78% [22]. By studying the thermal environment of railway stations, Japanese scholars have concluded that the temperature and the wind speed are main factors determining the thermal environment of the station where passengers stay for a short time. Furthermore, they put forward that ventilation is the most effective method to improve the thermal environment of the station. They even considered that a high wind speed can control the temperature between 18 °C and 32 °C in summer to ensure the same thermal comfort [23]. Thermal comfort ranges of buildings are very different from climates. Manual adjustments such as natural ventilation (opening windows in summer, ventilation, etc.) are more comfortable and energy-saving than mechanical methods [24–26]. In addition to objective analysis, people's subjective perception of the environment will also affect the evaluation of indoor comfort. For example, for wooden buildings, the use of environmental materials will increase the good impression of the internal environment [27,28]. Railway station buildings have a wider acceptable temperature range due to a shorter stay time than office buildings, but the design parameters of public buildings are still used in domestic design codes, which will cause energy waste [29].

Most studies of the light environment relied on subjective surveys. Light environment is influenced by the parameters including human activity, light intensity, and lighting design of buildings [30,31]. Results showed that the visual clarity, the color authenticity, the light and the shade contrast were the main factors influencing the light environment comfort in offices, and the corresponding weights were 29.55%, 17.59%, 16.77%, and 14.44%, respectively [32]. For the lighting environment of railway stations, the balance between natural lighting and energy consumption was studied. Glazing ratio was parabolic in relation to the load [33]. However, the high-speed railway station is much higher than the traditional railway station, with a larger glass enclosure structure to meet a better visual effect. From the perspective of energy saving and emission reduction, it is necessary to understand the current situation of the light environment in such buildings and to analyze its energy-saving potential.

Concentration of CO_2 in the building reflects the effect of ventilation. Many studies have shown that natural ventilation is able to meet the demand of indoor fresh air [34,35]. Natural ventilation was successfully used in some small office buildings [21,36,37]. While, for tall buildings, passive air infiltration will bring excessive energy consumption and damage the indoor thermal environment. Huang et al. [38–40] found that the thermal stratification in large single buildings such as gymnasiums was significantly higher than that in other public buildings due to a higher ceiling height (9–17 m) and multiple openings. This phenomenon is more serious in winter than in summer, and the load caused by the air infiltration accounts for 40–60% of the total load. Test results of CO_2 concentration in the waiting lounge of the station is about 900 ppm, which is higher than the 700 ppm stipulated in ASHARE (American Society of Heating Refrigerating and Airconditioning Engineer), but passengers still can accept it [14]. Compared to the former, the high-speed railway station with more openings and a higher ceiling height will have more obvious chimney effect in winter, which may increase the heating load in winter.

Current research results show that there is a huge difference between railway stations and other public buildings, and it is not suitable to use the evaluation method of public buildings for the railway stations. With the rapid development of the high-speed railway in China, a large amount of high-speed railway station buildings will be built in the future. However, designers and operational staff do not pay enough attention to the energy consumption and the indoor environment of such buildings. The main purpose of this study is to investigate the characteristics of energy consumption and the indoor environment of high-speed railway stations in China. Energy consumption is related to the design parameters of the station and the scale of the station building. Problems and suggestions of the high-speed railway station are proposed by comparing the previous results with the measured results. Studies of energy consumption characteristics and light-thermal environment characteristics of high-speed railway stations are beneficial to understand the energy consumption and environmental status of high-speed railway stations in China, to formulate relevant design specifications and to further improve thermal comfort of the station environment and to save energy.

2. Description of Energy Use Characteristics

2.1. Description of Data Source

In order to discover the current situation and characteristics of the energy consumption of high-speed railway stations, it is necessary to investigate the energy consumption data of high-speed railway stations of different scales in different climate regions for analysis and collation. According to the Design Standard for Energy Efficiency of Public Buildings (GB50189-2015) [41], five climates are considered, including severe cold (A), cold (B), cold winter and hot summer (C), warm winter and hot summer (D,) and temperate (E) (Figure 1). According to the Code for Design of Railway Station (TB-10100-2018) [42], of stations with a number of passengers dispatched during the peak hour, more than 5000 are large ones and within that, between 1000 and 5000 are medium-sized ones. In this paper, a total of 15 large and medium-sized stations are selected for measurements. The basic information for the selected railway stations is shown in Table 1. In this paper, the annual energy consumption, the passenger flow and the equipment load list of 15 high-speed railway stations in 2018 are provided by the operation department of the station, and the building information of the station is provided by the corresponding design company.

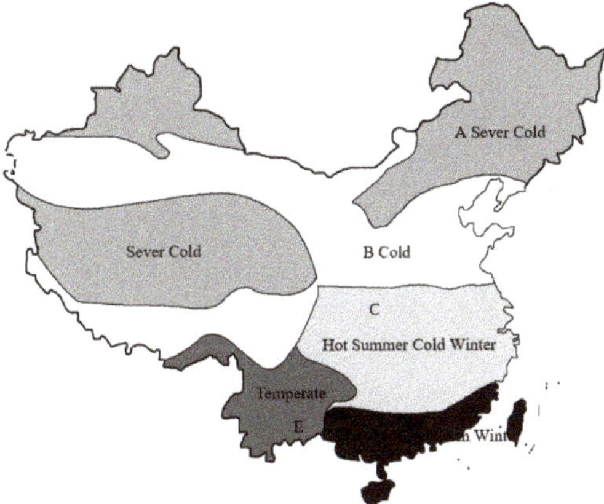

Figure 1. Five climates in China.

Table 1. General situation of stations.

Climate Zone	Station Number	Construction Area (10^4 m^2)	Annual Passenger Flow (10^4)	Station Scale	Area of Waiting Lounge (10^4 m^2)	Maximum Height (m)
A	A1	12.0	2675	Large	2.7	30
	A2	6.2	1500	Medium	1.7	18
B	B1	10.7	2500	Large	6.0	21
	B2	25.8	8546	Large	5.0	18
	B3	10.4	892	Medium	4.1	37
	B4	6.9	649	Medium	2.5	30
C	C1	24.0	4486	Large	7.8	21
	C2	28.2	6667	Large	8.1	33
	C3	10.8	4564	Large	4.0	25
	C4	11.0	2448	Large	3.4	18
D	D1	16.8	6796	Large	7.6	20
	D2	9.0	3650	Medium	4.3	20
	D3	6.9	680	Medium	NA [a]	NA
E	E1	12.0	858	Large	5.4	18.5
	E2	11.0	2448	Large	4.0	21

[a] NA, not available.

Energy sources used in railway station buildings mainly include electricity and natural gas. In order to facilitate the comparison, all kinds of energy are converted into electricity consumption. Energy consumption of the high-speed railway station comes from the data statistics of the station operation department. Annual and monthly building energy consumption of 2018 is read out by the meters of the distribution station. Energy consumption of items (including lighting, elevator, HVAC system, etc.) is separated by the equipment number and the energy use recorded. During the heating period, energy consumption is calculated by reading the boiler gas meter (Figure 2) or by heating costs.

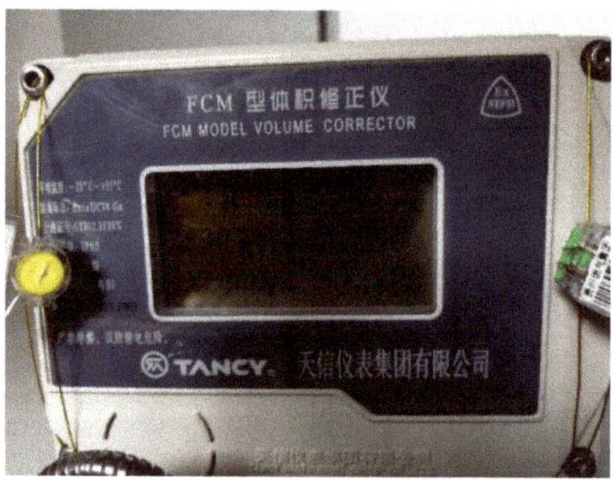

Figure 2. Gas use record of E2 station.

Most of the investigated high-speed railway stations are not equipped with the sub-item energy consumption measurement system. Annual energy consumptions of the building HVAC system and the heating system are obtained by the energy splitting method [43]. Monthly average power consumption in the transition period, annual energy consumption of the cooling system, and annual energy consumption of the heating system are calculated by Equations (1)–(3) [44,45], respectively:

$$E_{gd,j} = \frac{\sum_{i=1}^{M} E_{gd,i}}{M}, \tag{1}$$

$$E_c = E_{cz} - E_{gd,j} \times N_c, \tag{2}$$

$$E_n = E_{nz} - E_{gd,j} \times N_n, \tag{3}$$

where $E_{gd,j}$ is monthly average consumption in the transition period, kWh/month. $E_{gd,i}$ is the monthly energy consumption in the transition period, kWh. M is total number of months during the transition period. E_c is annual energy consumption of the air conditioning system, kWh. E_{cz} is total energy consumption in the cooling period, kWh. N_c is cooling month number. E_n is annual energy consumption of the heating system, kWh. E_{nz} is total energy consumption in the cooling period, kWh. N_n is heating month number.

Energy consumption of lighting system is calculated by Equation (4) through statistics of lights in each station:

$$E_L = \sum_{i}^{m} N_i \times P_i \times T_i \times D_i \times k_i \times 10^{-3}, \tag{4}$$

where E_L is lighting energy consumption, kWh. N_i is number of lights. P_i is power of light, W. T_i is average daily running time of lights, hour. D_i is number of days of light operation throughout the year. k_i is opening coefficient of lighting, the ratio of the actual number of lights to the total number. m is number of rooms.

2.2. Energy Consumption Characteristics of High-Speed Railway Stations

All types of energy are converted into equivalent power consumption according to the Classification and Presentation of Civil Building Energy Use (GB/T 34913-2017) [46]. Energy consumption of the high-speed railway station is different from that of the previous stations [29,42]. The annual total energy consumption of the high-speed railway stations ranges from 20×106 to 60×10^6 kWh. Figure 3 shows the annual energy use intensity (EUI) of different high-speed railway stations is between 115 and 470 kWh/(m^2·a). Results are 10–20% higher than the traditional railway stations [10,47], as the comparison in Table 2. EUI of different stations is related to the climate region and area of the station.

The total energy consumption of smaller stations (B3, D2) is lower, but the EUI is larger. The average EUI of the cooling system in each station is 35.12 kWh/(m^2·a), and the heating system is 56.93 kWh/(m^2·a). For the overall energy consumption of station buildings, the energy consumption in each station is quite different, but the change of energy consumption proportion in the same climate is very close. Most stations in regions D and E where the climate is warmer in winter do not have the heating system. Regions A and B mainly rely on gas boilers or municipal centralized heat for district heating. With a lower outside air temperature, the energy consumption of the heating system in cold the region accounts for 30–50%. In summer, the energy consumption of the cooling system in regions C and D accounts for 15–35% because of the large moisture load and the high outdoor temperature.

Table 2. Energy consumption data of different buildings.

	Annual	Cooling	Heating	Lighting	Others
High-speed station (kWh/(m^2·a))	117–450	25%	39%	13%	23%
Traditional station (kWh/(m^2·a))	101–269	20%	44%	18%	18%
Urban office building (kWh/(m^2·a))	30–107	35%	40%	15%	10%

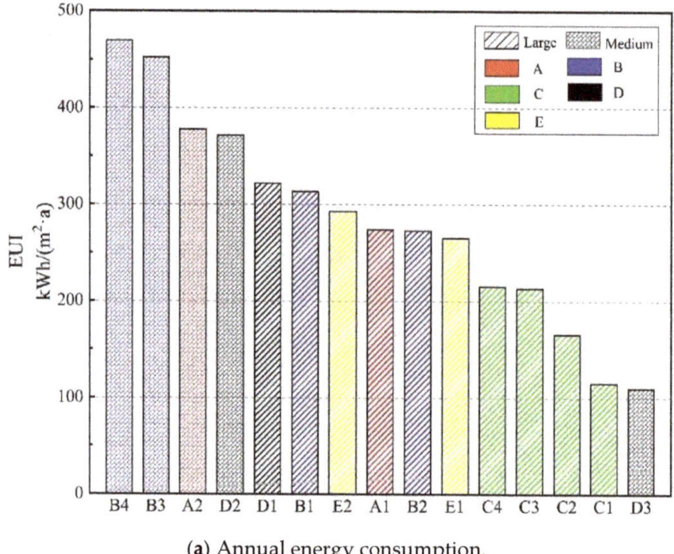

(**a**) Annual energy consumption.

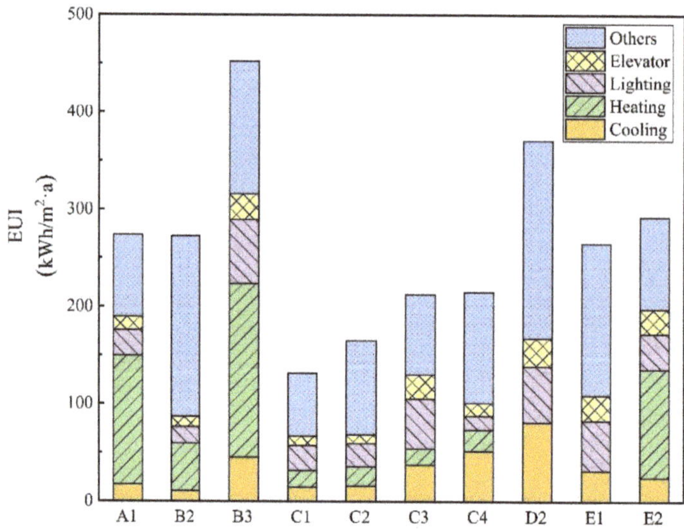

(**b**) The energy consumption proportion of each part of the system.

Figure 3. Statistical results of energy consumption of high-speed railway stations.

EUI in regions A and B is the largest, while EUI in regions C and E is the smallest. EUI in medium-sized stations is higher than that in large stations. As can be seen from the boxplot (Figure 4) made from the sample results in accordance with the climatic zone, EUI values can be roughly divided into high and low ranges. The main reason for the difference in climate interval is the heating method in winter. According to the classification of EUI of heating, the energy consumption in winter (111.5 kWh/(m²·a)) using the district heating is much higher than that of the station using non-district heating (Figure 5) (18.7 kWh/(m²·a)). Figure 6 shows the relation between EUI and station area by dividing the sample into two types, district heating and non-district heating. EUI of both stations is related to the number of the area, but the gap of EUI will increase with the area.

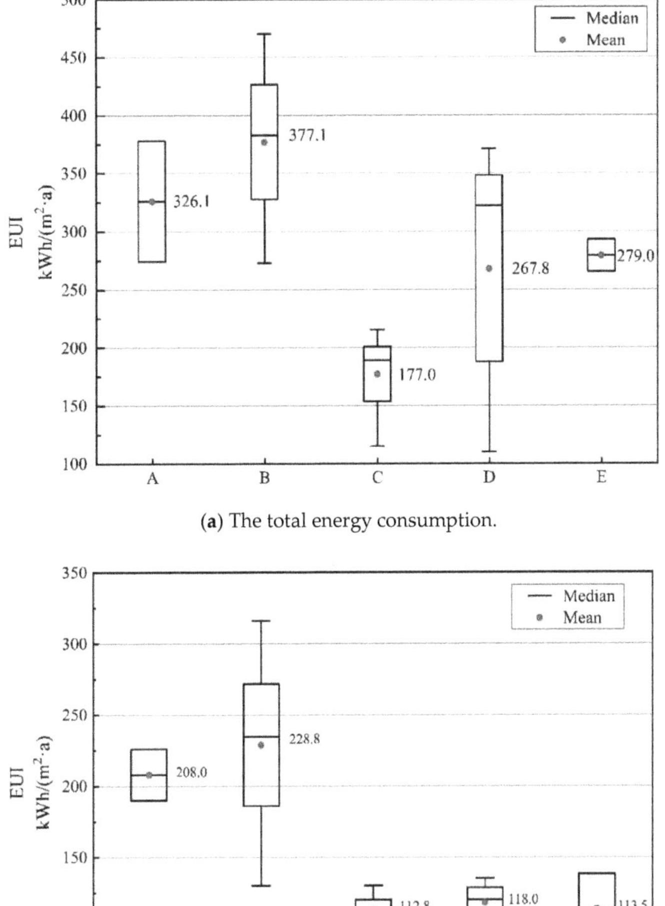

(**a**) The total energy consumption.

(**b**) HVAC (Heating Ventilation Air Conditioning), lighting, elevator system energy consumption.

Figure 4. Box plot of EUI (energy use intensity) of high-speed railway stations.

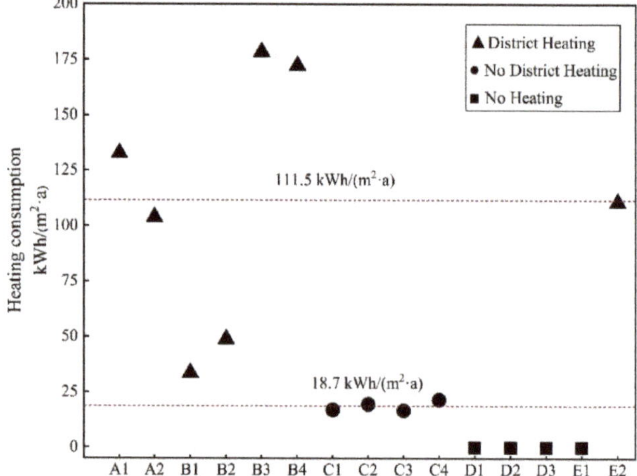

Figure 5. EUI of heating of high-speed railway stations.

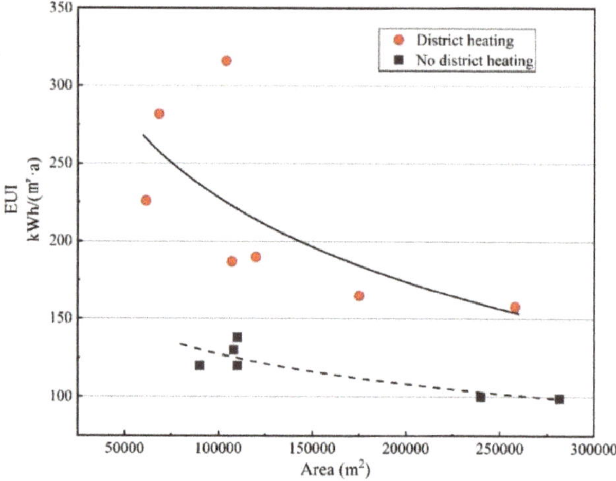

Figure 6. The relation between station area and EUI of different heating methods.

3. Environment Measurements

Energy consumption and passenger comfort are related to the quality of the environment. It mainly includes light comfort and lighting energy consumption, thermal comfort and HVAC system energy consumption, air quality and mechanical ventilation and infiltration air volume. Although people's subjective perception of the environment, such as the material of the structure and the design style of the building, can also affect the comfort of the building, it is not taken into account in this paper because there is almost no difference in the construction of the high-speed railway station. In this work, the indoor illumination, the thermal comfort, and the air infiltration of selected high-speed railway stations are measured. All the analyzes of the environmental quality of high-speed railway stations are based on the test results.

3.1. Error Analysis

As errors are often accompanied in the process of testing, in order to determine the accuracy of measurement results, error analysis of test data is needed. In the measurements, the errors of direct and indirect measurements of physical quantities are calculated by Equations (5) and (6), respectively 40. Parameters of the instrument required are shown in Table 3.

$$\gamma_i = \left(\frac{w_i^2}{3}\right)^{0.5},\tag{5}$$

$$\gamma_R = \left(\sum_{n=1}^{n}\gamma_i^2\right)^{0.5} = \left(\frac{1}{3}\sum_{n=1}^{n}w_i^2\right)^{0.5},\tag{6}$$

where γ_i is the ith direct measurement error of physical quantity. w_i is the ith direct measurement of physical quantity. γ_R errors in indirect measurements of physical quantities. n is the number of the direct measurement of physical quantities which is related to indirect measurement of physical quantities.

Table 3. Parameters of the instrument.

Name	Type	Parameter	Accuracy	Measurement Error
Hot wire anemometer	Testo-425	Wind speed	±0.03 m/s	0.017 m/s
Range finder	BS-GLM80	Distance	±0.002 m	0.001 m
Multifunctional Tester	Testo-435	Temperature, humidity and CO_2	±0.2 °C, ±1%, ±10 ppm	0.115 °C, 0.57%, 5.7 ppm
Luminometer	TES-1335	Illumination	±0.01 Lux	0.0057 Lux
Black-bulb thermometer	AZ8758	Black-bulb temperature	±0.2 °C	0.115 °C

3.2. Measurement Method

The high-speed railway station is mainly composed of the arrival hall, waiting lounge, offices, toilets, and equipment rooms. Among them, the waiting lounge and the arrival hall are the public areas for passengers, and in the waiting lounge, passengers stay for the longest time. At present, the glass enclosure structure is widely used in newly-built high-speed railway station buildings, which fully shows the advantages of natural lighting, whereas the building load may be increased [45]. Light environment is tested in the waiting lounge. The testing time starts from 8:00 to 18:00 during each day, with 2–3 days for each station. According to the space characteristics of the waiting lounge, the measuring points are evenly distributed. Take the station B3 for example, 15 test points as depicted in Figure 7 are placed to record the data every two hours using the luminometer TES-1335. The illuminance value of each measuring point in the waiting lounge is tested every two hours to calculate the variation rule of the average illuminance value in the high-speed railway station.

Measurements of thermal environment mainly focus on the waiting lounge and the entrance hall. Parameters including air temperature, humidity, wind speed, and average radiation temperature (black globe temperature) are tested. Thermal sensation is also affected by the outdoor temperature, so the outdoor temperature is tested at the same time. The activity intensity and the personnel density of passengers in different functional areas vary, and the thermal environment is different from each other. To measure the air temperature difference between the inside and the outside, four measuring points are placed outside the station and four in the entrance hall. Measuring points in the waiting lounge are the same as those in the light environment test. Wind speed measurements are placed at 1.5 m above the floor, as shown in Figure 8. All data are recorded every two hours. In addition, a questionnaire survey is conducted among passengers, including age, waiting time, thermal comfort, light comfort, and so on. Each station receives 100 questionnaires per day.

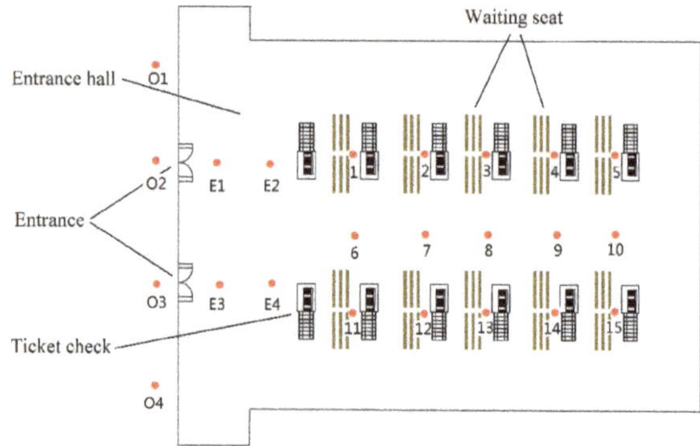

Figure 7. Layout of measuring points for B3.

Figure 8. Tests of waiting lounge for B3.

The air infiltration is measured by CO_2 concentration method, and the air change rate of the station is calculated by the difference of the CO_2 concentration between indoor and outdoor. CO_2 concentrations both in the waiting lounge and outside the station are measured by the Multifunctional Tester, and the location of the measuring point is the same as that in the thermal environment test. Each measuring point is tested every two hours. To ensure the accuracy of the data, each test lasts five minutes until the reading is stable.

3.3. Results

3.3.1. Light Environment

A large amount of glass enclosure is used as the enclosure structure of the high-speed railway station, as shown in Figure 9. The station glazing ratio is between 50–70%. Average illumination in most stations is much higher than the standard requirement of 100 Lux [48]. However, this does not mean that the good lighting effect is obtained, and the homogeneity needs to be considered. The use of natural lighting makes the illumination in the station vary greatly during the day. Figure 10 shows the change of illumination average value of different depth positions in B2 station during a summer day. It can be seen that the area affected by the natural lighting in B2 station accounts for 4/7 of the waiting room space of the station. The illumination in the middle of the station is relatively stable, which leads to the poor uniformity of illumination in the station. The main reason may be related to the regional radiation amount, the height of the station building, and the length of the lighting direction. Considering the maximum period of natural lighting (12:00), comparing the lighting homogeneity and the ratio of height to depth in different stations, it is found that there is a logarithmic relationship between them (Figure 11). Increasing the height can achieve better homogeneity but with the increasing energy consumption of HVAC. It is more feasible to use the shading or reduce the proportion of glass maintenance structure to alleviate the problem of uneven illumination.

(**a**) Waiting lounge of B3.

(**b**) Arrival hall of B1.

Figure 9. Glass curtain wall of high-speed railway station.

Passengers reflect that the daytime is brighter and the appropriate shading should be considered to improve the light comfort. The lighting control mode in the station is relatively simple and the response is slow with a certain time lag, which is also proved by the station staff. Normally, the glass curtain wall has poor thermal insulation and low air tightness. It is necessary to balance the relationship between natural lighting and energy consumption of the HVAC system to optimize the design.

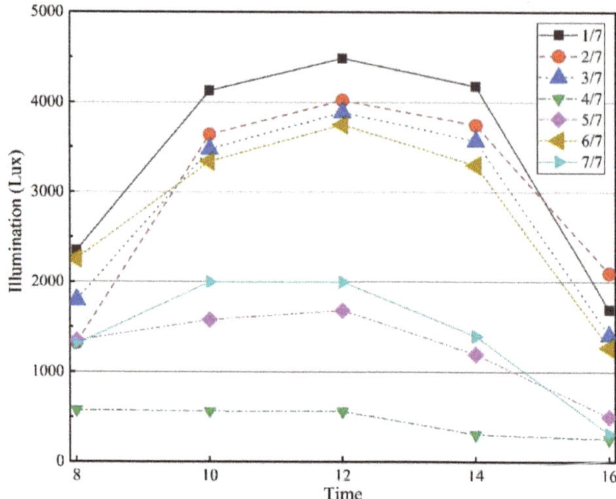

Figure 10. Mean of illumination of different depth position of B2 station.

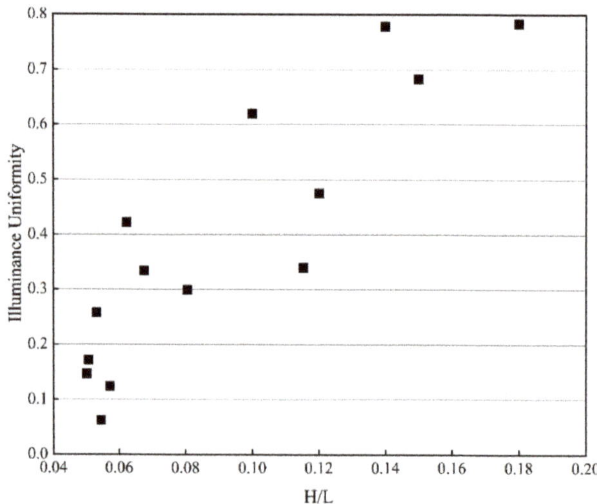

Figure 11. The relationship between illumination uniformity and station shape. H/L: The ratio between the height of the station and the length of the lighting direction. Illumination uniformity: The ratio of the lowest value of illumination to the average value.

3.3.2. Thermal Comfort and Thermal Environment

The measured indoor temperature, humidity, radiation temperature, wind speed, and outdoor temperature are shown in Tables 4 and 5. During the test, air conditioning systems are closed in the stations D1, D2, D3, and E1. Two hundred valid questionnaires are carried out in each station, including passenger age, gender, waiting habits, comfort, and other information.

Table 4. Winter test results.

Number	Indoor Temperature (°C)			Humidity (%)			Radiation Temperature (°C)			Outdoor Temperature (°C)			Wind Speed (m/s)		
	Max	Min	Mean	Max	Min	Mean	Max	Min	Mean	Max	Min	Mean	Max	Min	Mean
A1	22.0	16.1	18.5	24.4	14.1	18.8	22.1	17.2	19.4	4.2	−0.2	3.1	0.8	0.05	0.3
A2	22.5	11.6	18.7	12.9	6.5	8.1	22.1	11.6	18.4	−4.7	−5.6	−5.3	1.7	0.05	0.5
B1	17.5	10.6	15.3	25.4	14.8	17.2	17.4	4.8	14.9	1.5	0.2	1.0	1.2	0.01	0.4
B2	21.2	11.8	17.4	19.8	11.1	15.1	20.6	11.7	17.3	0.7	−3.3	−0.9	1.8	0.01	0.5
B3	20.2	13.6	17.7	21.8	9.9	20.3	19.4	11.7	12.2	5.7	1.7	3.6	1.0	0.01	0.2
B4	21.6	13.5	18.9	40.0	24.3	28.8	21.5	16.1	19.4	5.5	3.7	4.4	1.0	0.01	0.4
C1	20.8	12.0	16.1	40.4	23.5	31.2	19.3	10.8	15.8	6.9	2.8	4.8	1.1	0.02	0.3
C2	23.1	16.4	20.1	65.7	45.3	52.4	22.7	15.5	20.1	13.8	9.9	12.1	1.0	0.01	0.3
C4	20.8	12.0	16.1	40.4	23.5	31.2	19.3	10.8	15.8	6.9	2.8	4.8	1.1	0.02	0.3
D1	23.1	16.4	20.1	65.7	45.3	52.4	22.7	15.5	20.1	13.8	9.9	12.1	1.0	0.01	0.3
D2	22.0	20.5	21.4	67.7	61.6	64.6	22.4	20.5	21.5	21.1	18.9	20.1	0.5	0.01	0.2
D3	20.0	17.6	18.8	86.8	73.2	82.8	21.8	17.8	19.6	20.5	20.1	20.3	0.5	0.1	0.2
E1	15.4	13.0	14.0	70.5	62.4	65.8	15.0	11.7	13.6	12.5	11.6	11.9	0.8	0.1	0.3
E2	18.5	14.6	16.8	63.9	55.3	59.1	18.6	14.6	16.9	17.0	15.6	16.5	0.6	0.01	0.2

Table 5. Summer test results.

Number	Indoor Temperature (°C)			Humidity (%)			Radiation Temperature (°C)			Outdoor Temperature (°C)			Wind Speed (m/s)		
	Max	Min	Mean	Max	Min	Mean	Max	Min	Mean	Max	Min	Mean	Max	Min	Mean
A1	25.9	23.2	24.6	44.4	34.1	38.9	25.7	23.2	24.6	27.6	24.2	26.3	0.6	0.1	0.2
A2	25.5	21.5	24.1	32.9	30.1	31.5	25.7	22.5	24.2	27.3	24.6	25.2	1.5	0.1	0.5
B1	27.7	25.4	26.5	46.2	35.1	38.4	27.3	25.6	26.5	27.3	27.3	27.3	1.6	0.1	0.5
B2	29.2	24.1	26.3	40.1	31.1	35.2	30.2	24.5	27.0	30.6	27.5	29.6	1.2	0.0	0.3
B3	25.0	22.7	23.9	41.8	30.9	39.2	25.1	22.6	24.0	25.9	23.1	24.8	1.8	0.0	0.4
B4	25.9	24.1	24.9	55.5	44.3	48.9	25.7	24.1	25.0	27.1	24.0	25.4	1.5	0.1	0.5
C1	25.9	24.1	25.1	69.5	61.5	65.5	25.7	24.1	24.9	27.5	24.3	26.1	1.2	0.1	0.4
C2	27.8	23.2	24.5	70.1	59.8	63.9	27.8	23.5	24.6	27.3	26.3	25.3	1.6	0.1	0.6
C3	26.1	23.3	24.4	68.8	60.7	65.4	25.3	22.2	24.0	25.3	23.8	24.6	1.5	0.0	0.3
C4	25.5	23.8	24.8	65.4	54.0	61.3	25.9	24.6	25.3	24.9	24.3	24.7	0.9	0.1	0.2
D1	32.5	27.4	30.0	78.5	65.4	70.2	36.7	28.1	31.3	33.4	33.4	33.4	1.0	0.1	0.2
D2	29.7	26.1	28.0	82.5	70.2	78.2	29.9	26.7	28.3	37.2	37.2	37.2	1.1	0.0	0.3
D3	28.8	26.9	27.7	80.2	69.5	77.6	29.0	27.1	27.7	27.6	37.2	37.2	1.5	0.0	0.3
E1	27.0	22.2	24.8	75.5	62.4	70.1	26.4	22.7	24.6	27.4	23.3	25.3	0.9	0.0	0.1
E2	24.7	22.0	23.7	70.5	65.3	67.8	27.6	22.3	25.3	25.8	23.3	24.5	1.5	0.0	0.3
A1	25.9	23.2	24.6	44.4	34.1	38.9	25.7	23.2	24.6	27.6	24.2	26.3	0.6	0.1	0.2
A2	25.5	21.5	24.1	32.9	30.1	31.5	25.7	22.5	24.2	27.3	24.6	25.2	1.5	0.1	0.5
B1	27.7	25.4	26.5	46.2	35.1	38.4	27.3	25.6	26.5	27.3	27.3	27.3	1.6	0.1	0.5
B2	29.2	24.1	26.3	40.1	31.1	35.2	30.2	24.5	27.0	30.6	27.5	29.6	1.2	0.0	0.3

PMV (Predicted Mean Vote) ranges from hot (+3) to cold (−3) according to the ASHARE 7 scale [49].Through the survey of passenger thermal comfort satisfaction in winter, it is found that the majority of PMV are within −1~+1 as depicted in Figure 12a, indicating that the temperature in the waiting lounge can be accepted by most passengers. Most uncomfortable passengers feel that the station environment is cold in winter. Moreover, the thermal sensation of passengers in regions C and D is much colder than that in regions A and B, and the average temperature measured is lower as well. Main reasons are the high humidity in these climatic regions, the large opening of buildings and the lack of cold wind protection measures in winter. Because there is no need to consider the dehumidification in regions A and B, the windshield curtain wall and the glass house are installed in the entrance and the exit, where the insulation effect is improved.

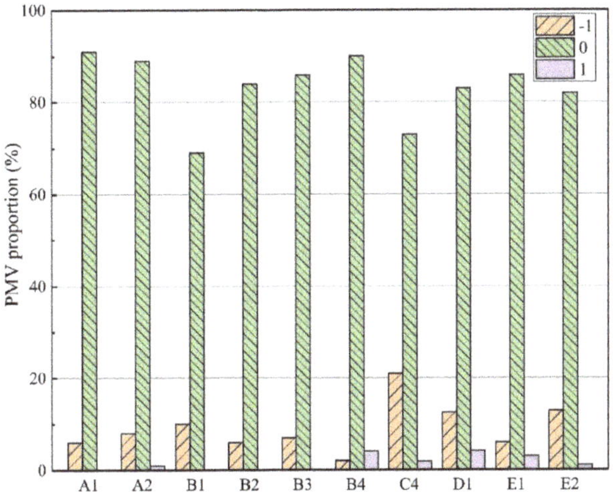

(**a**) PMV (Predicted Mean Vote) proportion of stations

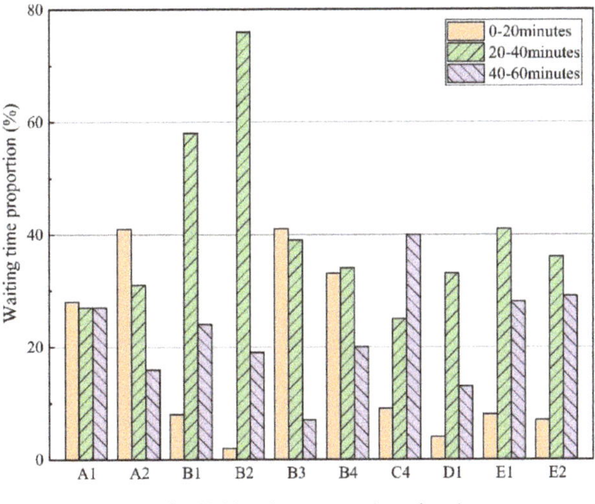

(**b**) Waiting time proportion of stations

Figure 12. PMV and waiting time proportion of stations.

Thermal sensation in the waiting lounge changes with the stay time. When passengers enter into the station, there is a kind of warm comfort due to the temperature difference between the indoor and the outdoor. As they stay longer, the comfort decreases. The average waiting time for passengers at traditional passenger stations is 68 min [50]. Questionnaire survey shows that the waiting time of passengers in high speed railway stations is shorter within 60 min. Very few passengers wait more than an hour. Figure 12b shows the proportion of passenger waiting time at different stations. The average passenger waiting time at all stations is about 40% within 20–40 min. As A2, B3, and B4 are medium stations, they have more short-distance trains, so the waiting time of passengers is shorter, accounting for 38% of passengers in 20 min. However, A1, B1, B2, D1, E1, and E2 are large stations with a high passenger flow, so the waiting time of passengers in 40–60 min reaches 25.4%. Therefore, passengers feel colder. The difference in sensation increases with the temperature difference between the indoor and the outdoor, and the thermal sensation tends to stabilize after about 40 min. Smaller stations have a higher proportion of passengers feeling comfortable due to the shorter waiting time. Regions A and B have a higher acceptance of the environment because the temperature difference between indoor and outdoor in winter is much greater than that in other regions.

According to ASHARE Standard 55-2017, the operative temperature based on Equation (7) is used for the analysis of thermal comfort in the waiting lounge, which considers the influence of air temperature and average radiation temperature on the human thermal sensation [51,52]:

$$t_0 = \alpha t_a + (1 - \alpha)t_r, \tag{7}$$

where t_0 is operative temperature, °C t_a is indoor air temperature, °C. t_r is average indoor radiation temperature, °C. α is factor related to the wind speed. When the indoor wind speed is 0–0.2 m/s, take 0.5, when the indoor wind speed is 0.2–0.6 m/s, take 0.6; when the indoor wind speed is 0.6–1.0 m/s, take 0.7.

According to the Reference [53], the mean number of thermal sensory votes (MTSV) is linearly regressed with the operative temperature (t_0). Thermal neutral temperature (T_0) is obtained when MTVS is 0. Table 6 gives the calculated thermal neutral temperature (T_0). In winter, passengers at station B4 feel comfortable and warm in the waiting area, and T_0 in the waiting lounges of other passenger stations is 0.3–3.2 °C higher than the average operative temperature, indicating that the thermal sensation in the waiting area is cold. This is consistent with the results of the thermal environment test.

Table 6. Thermal neutral temperature of high-speed railway stations.

	Station Number	A1	A2	B1	B2	B3	B4	C4	D1	E1	E2
Winter	T_0 (°C)	18.9	20.3	17.9	19.9	18.3	18.8	17.1	16.5	17.0	16.6
	Test temperature (°C)	18.5	18.6	15.1	17.3	17.5	19	13.9	16.8	16.2	15.2
Summer	T_0 (°C)	25.4	26.0	26.4	25.5	25.6	25.8	25.1	NA	26.0	26.3
	Test temperature (°C)	24.6	24.2	26.5	26.6	23.9	24.9	25.1	NA	25.0	25.2

The thermal neutral temperature of the waiting lounge varies greatly in different climate regions in winter. Although the temperature in the waiting lounge of most high-speed railway stations is lower than the design value (18 °C) in the code, most passengers can accept it. In winter, the thermal neutral temperature increases with the waiting time but decreases with the outdoor temperature. The highest thermal neutral temperature of 20.3 °C is found in region A, while the lowest value of 16.5 °C is found in region E. In addition, the waiting time of passengers at stations B3 and B4 is shorter than that at station B2, so the thermal neutral temperature is lower than that at station B2. The situation in summer is similar to that in winter. Therefore, when determining the design temperature, both the climate and the station scale should be considered. A single design parameter will cause the unnecessary energy waste.

According to ASHRAE [52] standard and ISO7730 [54], the regression equation is obtained by the polynomial regression with the operative temperature and the acceptable percentage. Taking the thermal environment with 80% of the passengers satisfied as the comfortable environment 36, the acceptable temperature range of passengers in waiting lounges of each station is concluded in Table 7.

Table 7. Acceptable temperature range in winter for waiting lounge (Unit: °C).

	A1	A2	B1	B2	B3	B4	C4	D1	E1	E2
Winter	17.9–19.8	18.1–21.5	16.4–20.0	17.4–21.7	17.5–19.9	17.9–20.9	15.1–18.8	15.4–18.5	15.2–18.6	15.6–18.0
Summer	24.6–26.5	25.5–26.5	25.0–26.5	25.4–26.4	24.5–26.5	25.3–26.5	25.0–26.3	25.5–27.0	25.6–26.7	25.7–26.9

Compared with other types of public buildings, the indoor acceptable operative temperature range of railway station is larger [55]. The acceptable temperature range in high-speed rail passenger stations is larger in winter than in summer. The average operative temperatures measured in stations B1, B2, C4, and E2 are lower than the acceptable range, so the passengers generally feel cold. The main reason is that the air conditioning system is not well maintained and a higher temperature in the station is needed. The average operative temperature measured by station B4 is close to the upper limit of the acceptable range, therefore, the station temperature appropriately can be reduced and energy use is saved.

3.3.3. Air Infiltration

Due to the high ceiling height (18–40 m) and the large number of openings in the ground floor, the air infiltration of the high-speed railway station is much larger than that of other public buildings. CO_2 concentration can be used as an index to evaluate the fresh air volume and the air quality in buildings. The permeable air volume of the stations is determined by measuring the average values of indoor and outdoor CO_2 concentrations in stations both in winter and in summer.

Ventilation standard proposed by ASHARE requires that the difference between the indoor and the outdoor CO_2 concentration should be less than 700 ppm, equivalent to the indoor CO_2 concentration of less than 1000 ppm [50], In this standard, the minimum fresh air volume for personnel is stipulated as well. The passenger flow of high-speed railway station is larger and it has more openings than other types of buildings. Test results show that the number of openings in winter in regions A and B is 4~8, and that in regions C and D it is 12~20. Table 8 shows the air change rate and the proportion of air infiltration to the load during typical days in winter and in summer for each station. Through interviews with staff, the introduction of mechanical fresh air in regions A and B in winter will do damage to the equipment because of very low outdoor temperature (−20~10 °C). Heating capacity of the heat pump in region C is insufficient in winter, and it cannot meet the fresh air load caused by plenty of mechanical fresh air. The main reason is the low temperature in stations in this area. Therefore, mechanical fresh air in winter is not used in these areas, while in regions D and E, the mechanical ventilation is opened.

Table 8. Test results of infiltration air.

		A1	A2	B1	B2	B3	B4	C1	C2	C4	D1	D2	E1	E2
Winter	Air change (h^{-1})	0.9	6.2	1.9	1.8	5.0	2.9	1.2	4.2	5.1	5.6	2.3	2.2	5.6
	Percent (%)	10	36	25	21	63	24	18	34	NA	NA	NA	22	NA
Summer	Air change (h^{-1})	0.7	2.0	1.9	1.2	1.3	2.9	0.5	3.0	2.7	1.3	1.2	4.1	2.8
	Percent (%)	10	19	10	29	22	34	48	39	40	NA	20	51	33

The measured CO_2 concentration in the waiting lounge of each station is very close to the outdoor value. Results show that the infiltration air volume in the station can meet or even exceed the indoor fresh air demand. Therefore, if the infiltration air volume cannot be controlled in the design, the mechanical fresh air can be adjusted to save energy. There is a logarithmic relationship between air infiltration volume and EUI (Figure 13), and the influence of osmotic wind on the station energy consumption in winter is greater than that in summer. Since the fresh air volume in the station is sufficient to meet the human body's needs (the number of ventilations in the specification is 1 h^{-1}), reducing the permeable air volume has become the most feasible energy saving method, and it is very effective to reduce the ceiling height or increase the resistance at the opening.

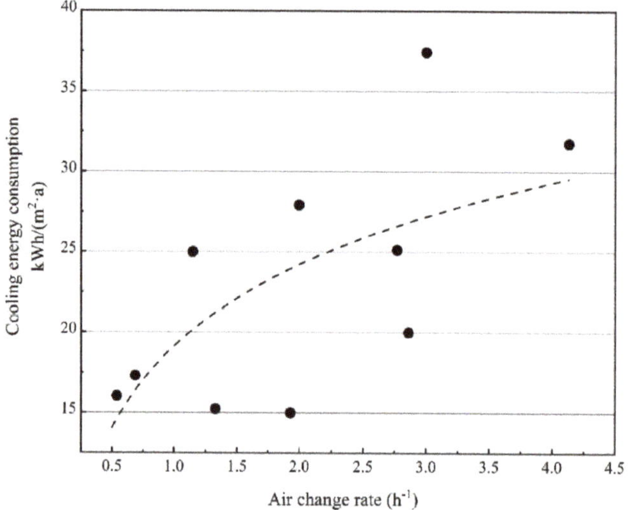

(**a**) Test results in the typical summer day.

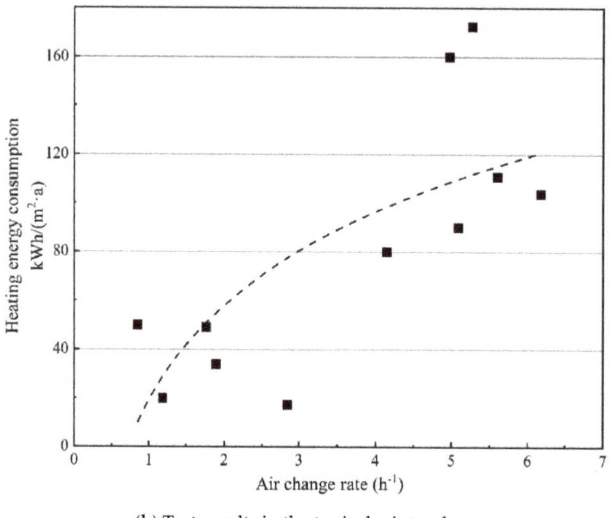

(**b**) Test results in the typical winter day.

Figure 13. The relationship between air infiltration and energy consumption.

4. Conclusions

This paper carries out some field measurements in 15 high-speed railway stations in five climatic regions in China. Some important findings are concluded as follows:

- The EUI of different high-speed railway stations is between 115 and 470 kWh/(m^2·a), which is related to the climate region and the area of the station. The average energy consumption of a station using district heating is 11.5 kWh/(m^2·a), while that of a station without district heating is 18.5 kWh/(m^2·a). EUI in stations decreases with the increase of area, and this trend is more obvious in stations using central heating.

- The use of a large number of glass maintenance structures makes the illumination in the station vary greatly during the day. The average illumination in most stations is much higher than the standard requirement of 100 Lux. However, this does not mean that the good lighting effect is obtained, and the distribution needs to be considered. The lighting evenness and the height of the station are related to the depth of the lighting direction. Increasing the height can achieve better evenness but with the increasing energy consumption of HVAC.

- The thermal neutral temperature of the waiting lounge varies greatly in different climate regions. Most passengers can accept the indoor temperatures below 18 °C or lower, which is related to the outdoor temperature. The main reason is that the short waiting time (40–60 min) of passengers and the requirements of thermal environment are different from those of office buildings. The acceptable temperature range in high-speed railway stations is larger in winter than in summer.

- The measured CO_2 concentration in the waiting lounge of each station is very close to the outdoor value. Results show that the infiltration air volume in the station can meet or even exceed the indoor fresh air demand. Maybe the mechanical fresh air can be canceled to save energy. There is a logarithmic relationship between the infiltration air volume and the energy consumption of the station and influence of infiltration air on energy consumption in winter is greater than that in summer.

In summary, compared with traditional railway stations, high-speed railway stations have higher energy consumption due to more openings, higher ceiling height, and a large number of glass maintenance structures. Passengers in high-speed trains also have different waiting habits, so their thermal comfort needs are also different. The high-speed railway station has considerable energy saving potential in the utilization of air infiltration, uniform lighting, and indoor temperature control. In addition, an efficient energy management platform is very important for the construction of high-speed railway stations.

Author Contributions: Conceptualization, B.Q. and H.B.; methodology, B.Q.; validation, B.L., T.Y.; formal analysis, B.Q.; investigation, B.Q.; data curation, B.Q.; writing—original draft preparation, B.Q.; writing—review and editing, B.L. H.B. and T.Y. All authors have read and agreed to the published version of the manuscript.

Funding: This research was funded by "National Key R&D Program of China" grant number [No. 2018YFC0705000], and The APC was funded by [No. 2018YFC0705000].

Acknowledgments: This work is supported by the "National Key R&D Program of China" (No. 2018YFC0705000).

References

1. Omer, A.M. Energy, environment and sustainable development. *Renew. Sustain. Energy Rev.* **2008**, *12*, 2265–2300. [CrossRef]

2. Perez-Lombard, L.; Ortiz, J.; Pout, C. A review on buildings energy consumption information. *Energy Build.* **2008**, *40*, 394–398. [CrossRef]

3. Jiang, Y. *Annual Report on China Building Energy Efficiency*; Building Energy Conservation Research Center of Tsinghua University; China Architecture & Building Press: Beijing, China, 2018. (In Chinese)

4. Xiao, H.; Wei, Q.; Jiang, Y. The reality and statistical distribution of energy consumption in office buildings in China. *Energy Build.* **2012**, *50*, 259–265. [CrossRef]

5. Bracke, W.; Delghust, M.; Laverge, J.; Janssens, A. Building energy performance: Sphere area as a fair normalization concept. *Build. Res. Inf.* **2019**, *47*, 549–566. [CrossRef]

6. Zhou, N.; Lin, J. The reality and future scenarios of commercial building energy consumption in China. *Energy Build.* **2008**, *40*, 2121–2127. [CrossRef]

7. Doctor-Pingel, M.; Vardhan, V.; Manu, S.; Brager, G.; Rawal, R. A study of indoor thermal parameters for naturally ventilated occupied buildings in the warm-humid climate of southern India. *Build. Environ.* **2019**, *151*, 1–14. [CrossRef]

8. Hoyt, T.; Arens, E.; Zhang, H. Extending air temperature setpoints: Simulated energy savings and design considerations for new and retrofit buildings. *Build. Environ.* **2015**, *88*, 89–96. [CrossRef]

9. Nicol, J.F.; Humphreys, M.A. Adaptive thermal comfort and sustainable thermal standards for buildings. *Energy Build.* **2002**, *34*, 563–572. [CrossRef]

10. Song, L.; Wang, Y.; Li, X. Energy performance and environmental quality of typical railway passenger stations in northern China. *Indoor Built Environ.* **2016**, *27*, 296–307. [CrossRef]

11. Yan, H.; Wang, K. 2011 National Railway Passenger Station Energy Consumption Special Investigation Situation Analysis. *Railw. Econ. Res.* **2012**, *5*, 8–13.

12. Wang, L.; Li, Q.; Dang, R.; Liu, G.; Liu, H. Comfort-Based Energy Efficiency Reconstruction of Waiting-room of Railway Station. *Build. Energy Effic.* **2014**, *3*, 97–102.

13. Yang, L.; Xia, J. Case Study of Space Cooling and Heating Energy Demand of a High-speed Railway Station in China. *Procedia Eng.* **2015**, *121*, 1887–1893. [CrossRef]

14. Ma, W.W.; Liu, X.Y.; Zhou, C.Q. Research on the waiting time of passengers and escalator energy consumption at the railway station. *Energy Build.* **2009**, *41*, 1313–1318. [CrossRef]

15. Shi, L.; Xie, M.J. Optimal Simulation Analysis of Day lighting Design in New Guangzhou Railway Station. In *Advanced Information Technology in Education*; Thaung, K.S., Ed.; Springer: Berlin/Heidelberg, Germany, 2012; pp. 197–206.

16. Brenna, M.; Dolara, A.; Foiadelli, F.; Leva, S.; Longo, M.; Zaninelli, D.; Casiraghi, F.M. Power quality analysis of LED lighting system for railway applications. In Proceedings of the IEEE International Conference on Harmonics & Quality of Power, Bucharest, Romania, 25–28 May 2014.

17. Xiao, J.; Liu, C.J. Lighting Control and Its Power Management in Railway Passenger Station. In Proceedings of the 2009 Asia-Pacific Power and Energy Engineering Conference, Wuhan, China, 28–30 March 2009.

18. Sengor, I.; Kilickiran, H.C.; Akdemir, H.; Kekezoglu, B.; Erdinc, O.; Catalao, J.P.S. Energy Management of a Smart Railway Station Considering Regenerative Braking and Stochastic Behaviour of ESS and PV Generation. *IEEE Trans. Sustain. Energy* **2018**, *9*, 1041–1050. [CrossRef]

19. Dounis, A.I.; Caraiscos, C. Advanced control systems engineering for energy and comfort management in a building environment—A review. *Renew. Sustain. Energy Rev.* **2009**, *13*, 1246–1261. [CrossRef]

20. Arens, E.; Humphreys, M.A.; de Dear, R.; Zhang, H. Are 'class A' temperature requirements realistic or desirable? *Build. Environ.* **2010**, *45*, 4–10. [CrossRef]

21. Tong, Z.; Chen, Y.; Malkawi, A.; Liu, Z.; Freeman, R.B. Energy saving potential of natural ventilation in China: The impact of ambient air pollution. *Appl. Energy* **2016**, *179*, 660–668. [CrossRef]

22. Kalz, D.E.; Pfafferott, J. *Thermal Indoor Environment, Thermal Comfort and Energy-Efficient Cooling of Nonresidential Buildings*; Springer: New York, NY, USA, 2014; pp. 15–20.

23. Nakano, J.; Tanabe, S.; Morii, T.; Uruno, M.; Goto, Y.; Sakamoto, K. *Field Study on Optimum Thermal Environmental Control of Train Stations*; The Society of Heating, Air-Conditioning Sanitary Engineers of Japan: Tokyo, Japan, 2006.

24. Kwong, Q.J.; Adam, N.M.; Sahari, B.B. Thermal comfort assessment and potential for energy efficiency enhancement in modern tropical buildings: A review. *Energy Build.* **2014**, *68*, 547–557. [CrossRef]

25. Boerstra, A.C.; te Kulve, M.; Toftum, J.; Loomans, M.G.; Olesen, B.W.; Hensen, J.L. Comfort and performance impact of personal control over thermal environment in summer: Results from a laboratory study. *Build. Environ.* **2015**, *87*, 315–326. [CrossRef]

26. del Ferraro, S.; Iavicoli, S.; Russo, S.; Molinaro, V. A field study on thermal comfort in an Italian hospital considering differences in gender and age. *Appl. Ergon.* **2015**, *50*, 177–184. [CrossRef]

27. Caniato, M.; Gasparella, A. Discriminating People's Attitude towards Building Physical Features in Sustainable and Conventional Buildings. *Energies* **2019**, *12*, 1429. [CrossRef]

28. Sant'Anna, D.O.; Dos Santos, P.H.; Vianna, N.S.; Romero, M.A. Indoor environmental quality perception and users' satisfaction of conventional and green buildings in Brazil. *Sustain. Cities Soc.* **2018**, *43*, 95–110. [CrossRef]

29. Huang, M.; Lin, Y. Thermal comfort of railway station's waiting room in severe cold regions of China. *Sustain. Energy Build.* **2017**, *134*, 749–756. [CrossRef]

30. Kawamoto, R.; Umemiya, N.; Atsuta, H.; Cie, T.S. *Relation between Light Environment Evaluation and Personal Attributes*; C I E Central Bureau: Vienna, Austria, 2011.

31. Liu, Y.; Yu, G.; Zheng, S.W. Gymnasium Natural Light Environment Optimization Design Oriented by Environment-behavior Studies. In *Advanced Materials Research*; Albahar, S.K., Zhao, J.Y., Eds.; Trans Tech Publications: Zurich, Switzerland, 2014; pp. 223–227.

32. Durak, A.; Olgunturk, N.C.; Yener, C.; Guvenc, D.; Gurcinar, Y. Impact of lighting arrangements and illuminances on different impressions of a room. *Build. Environ.* **2007**, *42*, 3476–3482. [CrossRef]

33. Li, D.H.W.; Lam, J.C.; Wong, S.L. Daylighting and its implications to overall thermal transfer value (OTTV) determinations. *Energy* **2002**, *27*, 991–1008. [CrossRef]

34. Ai, Z.T.; Mak, C.M.; Cui, D.J.; Xue, P. Ventilation of air-conditioned residential buildings: A case study in Hong Kong. *Energy Build.* **2016**, *127*, 116–127. [CrossRef]

35. Caciolo, M.; Cui, S.Q.; Stabat, P.; Marchio, D. Development of a new correlation for single-sided natural ventilation adapted to leeward conditions. *Energy Build.* **2013**, *60*, 372–382. [CrossRef]

36. de Dear, R.J.; Brager, G.S. Thermal comfort in naturally ventilated buildings: Revisions to ASHRAE Standard 55. *Energy Build.* **2002**, *34*, 549–561. [CrossRef]

37. Ai, Z.T.; Mak, C.M.; Cui, D.J. On-site measurements of ventilation performance and indoor air quality in naturally ventilated high-rise residential buildings in Hong Kong. *Indoor Built Environ.* **2015**, *24*, 214–224. [CrossRef]

38. Huang, C.; Zou, Z.; Li, M.; Wang, X.; Li, W.; Huang, W.; Yang, J.; Xiao, X. Measurements of indoor thermal environment and energy analysis in a large space building in typical seasons. *Build. Environ.* **2007**, *42*, 1869–1877. [CrossRef]

39. Said, M.N.A.; Macdonald, R.A.; Durrant, G.C. Measurement of thermal stratification in large single-cell buildings. *Energy Build.* **1996**, *24*, 105–115. [CrossRef]

40. Taraldsen, G. On the ISO Guide to the Expression of Uncertainty in Measurements. *J. Radioanal. Nucl. Chem.* **1995**.

41. *Design Stand for Energy Efficiency of Public Buildings*; GB50189-2015; China Architecture & Building Press: Beijing, China, 2015. (In Chinese)

42. *Code for Design of Railway Passenger Station*; TB 10100-2018; China Railway Publishing House: Beijing, China, 2018. (In Chinese)

43. Ma, W.; Li, L.; He, S.; Cheng, J.; Huang, G.; Zhou, C.Q. Influencing Factors on Energy Consumption of Air Conditioning System in Railway Passenger Station Based on Orthogonal Experiment. In Proceedings of the Second International Conference on Intelligent System Design & Engineering Application, Sanya, China, 6–7 January 2012.

44. Wu, Z.; Huang, K.; Wei, Y. Energy Consumption Division of HVAC System in Building Energy Audits. In *Building Energy Efficiency*; Shenyang, China, 2012.

45. Bouden, C. Influence of glass curtain walls on the building thermal energy consumption under Tunisian climatic conditions: The case of administrative buildings. *Renew. Energy* **2007**, *32*, 141–156. [CrossRef]

46. *Classification and Presentation of Civil Building Energy Use*; GB T34913-2017; China Quality and Standards Publishing: Beijing, China, 2017. (In Chinese)

47. Liu, X. *Research on Air-Conditioning Energy Consumption at the Passenger Railway Station in Summer*; Central South University: Changsha, China, 2010. (In Chinese)

48. *Code for Design of Railway Passenger Station Buildings*; GB 50266-2007; Standardization Administration of China (SAC): Beijing, China, 2007. (In Chinese)

49. Enescu, D. A review of thermal comfort models and indicators for indoor environments. *Renew. Sustain. Energy Rev.* **2017**, *79*, 1353–1379. [CrossRef]

50. Liu, X.; Ma, W. Research of Waiting Time of Passengers at Railway Stations. *J. China Railw. Soc.* **2009**, *31*, 104–107. (In Chinese)

51. Concepts, R. *Thermal Environmental Conditions for Human Occupancy*; ASHRAE Standard: Atlanta, GA, USA, 2014.

52. *Thermal Environmental Conditions for Human Occupancy*; ASHARE 55-2017; American Society of Heating, Refrigerating and Air-Conditioning Engineers: Atlanta, GA, USA, 2017.

53. Candido, C.; de Dear, R.J.; Lamberts, R.; Bittencourt, L. Air movement acceptability limits and thermal comfort in Brazil's hot humid climate zone. *Build. Environ.* **2010**, *45*, 222–229. [CrossRef]
54. *Moderate Thermal Environment-Determination of the PMV and PPD Indices and Specification of the Condition for Thermal Comfort*; ISO 7730; International Standard Organization: Copenhagen, Denmark, 2016.
55. Peeters, L.; de Dear, R.; Hensen, J.; D'haeseleer, W. Thermal comfort in residential buildings: Comfort values and scales for building energy simulation. *Appl. Energy* **2009**, *86*, 772–780. [CrossRef]

Article

Analysis of the Effect of Using External Venetian Blinds on the Thermal Comfort of Users of Highly Glazed Office Rooms in a Transition Season of Temperate Climate—Case Study

Małgorzata Fedorczak-Cisak *, Katarzyna Nowak and Marcin Furtak

Faculty of Civil Engineering, Cracow University of Technology, 31-155 Cracow, Poland;
knowak@pk.edu.pl (K.N.); mfurtak@pk.edu.pl (M.F.)
* Correspondence: mfedorczak-cisak@pk.edu.pl; Tel.: +48-696-046-050

Received: 24 October 2019; Accepted: 18 December 2019; Published: 23 December 2019

Abstract: Improving the energy efficiency of buildings is among the most urgent social development tasks due to the scale of energy consumption in this industry. At the same time, it is essential to meet high requirements for indoor environmental quality and thermal comfort. The issue of overheating is most often analysed in summer but it also occurs in transition seasons, when the cooling systems do not operate. The paper attempts to evaluate the effectiveness of external mobile shading elements on the microclimate of rooms with large glazed areas in the transition season. Passive solutions, such as shading elements, which limit the increase of indoor temperature, do not always allow the acquisition and maintenance of comfortable solutions for the duration of the season, as demonstrated by the authors. Temporary cooling of the rooms may be necessary to maintain comfortable conditions for the users, or other solutions should be devised to improve comfort (e.g., reduction of clothing insulation characteristics). The novelty of the study consists in the analysis of comfort in a "nearly zero energy consumption" building (NZEB) during a period not analyzed by other scientists. This is a transition period during which heating/cooling systems do not operate. The research task set by the authors involved the assessment of the possibility to reduce office space overheating in the transition season (spring) by using external shading equipment in rooms with large glazed areas. An additional research task aimed at checking the extent to which user behaviour, such as reduction in clothing insulation characteristics, can improve comfort in overheated rooms. The results of the tests reveal that the difference in the ambient air temperature between a room with external venetian blinds and an identical room with no venetian blinds in the transition season, i.e., from 27 March to 6 April 2017, ranged from 12.3 to 2.1 °C. The use of a shading system (external venetian blinds positioned at an angle of 45°) reduced the number of discomfort hours by 92% (during working hours) compared to the room without external venetian blinds. A reduction in the thermal insulation of the clothes worn by people working in the room with no venetian blinds helped to reduce the number of discomfort hours by 31%.

Keywords: thermal comfort; overheating; transition seasons

---

1. Introduction

The policy of the European Union obliges member states to introduce a new standard of nearly zero energy buildings [1,2]. The implementation of the directive on the energy characteristics of buildings translates onto requirements for buildings set out in the Technical and Building Conditions [3]. Passive buildings have also become more popular in Europe [4,5]. Both standards of buildings are characterised by very low energy demand. Designing such buildings requires wide knowledge of the discipline [6]. A number of newly designed office and public buildings are characterised by large glazed areas. Despite selecting the correct insulation characteristics for the building structure elements, excess glazing generates large and undesired solar gain especially in summer. As a result of such design solutions, uncomfortable working conditions are created. Therefore, it seems necessary to estimate thermal comfort conditions at each work station in the period of time when overheating occurs in the building. Proposals for system (need for cooling), architectural and building solutions aimed at the reduction of overheating should result from such analyses. The issue of creating and maintaining indoor conditions comfortable for users has already been described in the literature [7,8]. Fanger proposed that the subjective thermal comfort sensation of the users of rooms should be identified with PMV (predicted mean vote) and PPD (predicted percentage of dissatisfied) assessment indices. The method developed by Fanger was incorporated into European standards concerning the thermal comfort of rooms [9–11]. The need to reduce energy consumption for heating and cooling purposes, which has been increasing recently, encourages designers and researchers to carry out more exact analyses for ensuring thermal comfort. Studies are directed towards improving the comfort model [12,13] through a number of experimental tests in laboratories, as well as in real-life objects [14,15]. Many researchers include thermal comfort into indoor environment quality models, additionally taking into account acoustic and lighting comfort and proposing modifications [16–19].

Summing up, low energy consumption must not be the only design criterion. The indoor microclimate is a combined effect of design, building and use of specific rooms. Human activity, clothing insulation and environmental parameters, such as ambient air temperature, mean radiation temperature, air flow rate and relative humidity, affect thermal comfort. Thermal comfort assessment is based on such indices as PMV and PPD.

The paper describes the influence of external shading elements on thermal comfort in premises with large glazed areas. Energy-efficient windows, which are nowadays regularly installed in buildings, are reaching the point where, on an annual basis, heat gains through a window exceed losses. The architecture of contemporary public and office buildings is characterised by highly glazed facades. Even in temperate climate zones, which is where Poland is situated, the solution generates intensive solar gain in the rooms, especially in the southern and western façades. The study of indoor comfort and air quality has been undertaken by a number of research teams in different climate zones [20–22]. The studies have covered different categories of buildings: residential, office, exhibition and entertainment facilities [23–25]. The scope of the studies analysed in the paper has been limited to office premises in buildings with glazed facades. Overheating is a major problem, which occurs in this type of premises. A number of papers [26–28] quote algorithms and guidelines pertaining to the glazing area size depending on material and construction solutions, and location. Creative vision and visual effect tend to be more important for architecture designs. Designers and investors often neglect researchers' recommendations to include glazed elements in small areas of the façades. That is why the purpose of this paper is to evaluate the effectiveness of external mobile shading elements for rooms with large glazed areas. The results presented in [29] showed that the use of shading devices is of great importance in cities, where temperatures and the intensity of solar radiation are high, but it also helps to reduce the demand for cooling in locations of temperate maritime climate. The authors of papers [30,31] analysed the impact of selected shading devices on the heat and lighting characteristics of rooms. Many of the studies are purely simulations. They are usually carried out for the summer or winter periods [32]. In transition seasons in temperate climates there are more cloudless days with a high value of direct solar radiation intensity. The issue of the overheating of rooms in these seasons is not often analysed in the literature.

In Poland's climate zone, energy for heating purposes in buildings with well-insulated structure is supplied from mid-October to mid-March. This is the period when it may be necessary to cool rooms on sunny days. Typically, the cooling system operates from May to September. Simulation analyses and design calculations for the period are used as the basis for selecting the parameters of the outer casing of building partitions and installation systems.

Researchers studying thermal comfort in buildings typically focus on the summer and winter periods when heating and cooling systems are turned on. The novelty of the authors of this article is the analysis of the transition period when the heating or cooling systems of the room do not work. An additional value is the analysis of thermal comfort in a building with "almost zero energy consumption" (NZEB). NZEB buildings comply with the requirements of Directive 2010/31/EU and will be a standard in Europe from 1 January 2021. The article makes a valuable contribution to the design and use of this type of buildings, so that, in addition to the low energy consumption required by regulations, also achieving comfort of use.

Studies on thermal comfort in this article are connected with a transition period in the climate in Poland (spring). The analyzed time is the turn of March and April. At this time, at noon, the solar altitude of the Sun is 38.04° to 49°. This not-too-large angle of the Sun's height allows for intense penetration of solar radiation into the interior of the room. During this period, heating and cooling systems usually do not work. Most solar energy for the analyzed location is generated in July, June and May. Most thermal comfort tests in buildings are related to summer (cooling) or winter (heating) periods. In the transition periods (spring, autumn) it is assumed for a standard construction to avoid heating or cooling the rooms.

The NZEB building was used for analysis, which by definition should be characterized by good insulation of external partitions, adequate protection against overheating, and a correspondingly low value of energy used for heating and cooling. The authors wanted to show that with modern buildings, with a well-insulated block, strongly glazed passive solutions, limiting the temperature increase in the room, such as shading elements, even for a transitional period may not be sufficient to obtain and maintain comfortable conditions throughout the period, which the article showed the authors. The rooms analyzed were characterized by a large ratio of the glazed area to the floor area of 5.63. The tested rooms are an example of the current solutions of office buildings, characterized by high heat capacity of ceilings, light internal walls and large glazed surfaces.

Avoiding the use of external shading devices due to investment costs results in obtaining thermal conditions uncomfortable for users or the necessity to cool rooms by up to 10 K.

An additional goal of the research presented in the article was to indicate that in the case of objects with a large area of external glazing, the need for cooling should be included in the calculation of energy demand also for spring months. This, of course, is associated with an increase in energy demand indicators for the building and an increase in operating costs.

Tests were carried out in an experimental laboratory building, which serves a public function. The building in which the tests were carried out is situated in the centre of Cracow, among city centre structures. It is a passive technology building, which meets the requirements for nearly zero energy buildings (NZEB) in Poland. It is also an experimental building, designed for performing energy efficiency tests.

The research questions posed by the authors for the studied case are the following:

- How can solar gain affect thermal comfort conditions in highly glazed rooms in the transition season of temperate climates?
- Will the use of external shading devices in highly glazed rooms help to maintain conditions comfortable for users?
- What is the degree of thermal comfort improvement owing to user behaviour involving reduction in clothing insulation characteristics?

2. Materials and Methods

Tests on thermal comfort were carried out in an experimental building of the Krakow University of Technology—Małopolskie Laboratorium Budownictwa Energooszczędnego (MLBE—Energy-Efficient Building Laboratory of Lesser Poland). The MLBE building is dedicated to such experiments. The tests were performed in rooms with the same area 37.2 m^2, on the second and third floors. The location of rooms P1.06 and P2.04 is shown in Figure 1. The external façade of the selected MLBE rooms was southward and westward-oriented. This is a glazed façade with a total area of 26.20 m^2 in each room. Room P2.04 had external mobile shading venetian blinds fitted. Room P1.06 had no shading system (Figure 1).

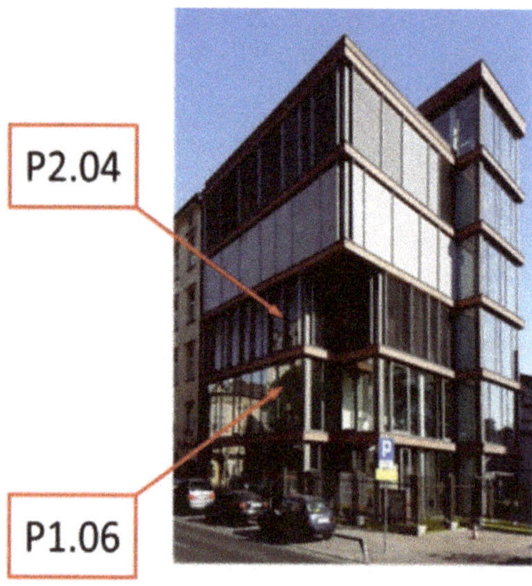

Figure 1. Location of test room P2.04 (with external venetian blinds) and P1.06 (with no venetian blinds).

The rooms selected for the tests represent typical office premises in office buildings of contemporary design.

The MLBE building was designed to meet the requirements of thermal protection for passive buildings [4]. The experimental MLBE building fulfils the Polish requirements for NZEBs presented in [3]. The thermal insulation parameters of the MLBE building and of the external partitions of passive buildings, and the Polish thermal insulation requirements included in [3] are presented in Table 1.

Table 1. Heat-transfer coefficient U W/(m^2·K) for Energy-Efficient Building Laboratory of Lesser Poland (MLBE) building, requirements according to Polish regulations [3] and requirements for passive buildings [4].

Type of Partition	MLBE Building U Parameters W/m^{2}K	Effective Requirements in Poland WT2017	Requirements for NZEB Buildings in Poland (Since 2021) WT2021	Requirements for Passive Buildings
External walls:	0.11	0.23	0.20	0.15
Roofs and floors:	0.12	0.18	0.15	0.15
Floor on the ground	0.11	0.30	0.30	0.15
Windows	0.8 $g = 0.6$ [-]	1.10	0.90	0.80

The experimental MLBE building, where tests were performed for the purpose of the study, was divided into 14 independent heating and cooling zones. This means that every zone can be heated or cooled independently. The MLBE building's control system was used to stabilise the conditions in the rooms adjacent to the test rooms. During the experiment, the control system was designed so that a temperature of +20 °C was maintained in the rooms adjacent to the test rooms.

The thermal comfort test presented in the paper was carried out for the transition season when the heating/cooling systems in the rooms were not in operation. The representative period lasted for 11 representative days, between 27 March and 6 April 2017. The representative period was selected with regard to the greatest number of cloudless days with direct solar radiation.

The air was supplied by an Air Handling Unit (AHU) with a recuperator with 90% heat recovery (Figure 2a). The quantity of the air supplied by a system of intake ventilators (Figure 2b) was constant and amounted to 25 m^3 per hour (assuming that only one user was present). The supplied air temperature was +18 °C.

(a)   (b)

Figure 2. (**a**) View of the Air Handling Unit (AHU) with a recuperator with 90% efficiency. (**b**) The view of the intake vent grille in room P1.06; the grille is located in the central part of the ceiling.

The test was carried out using two sets of sensors for thermal comfort tests. The set of sensors placed in room P2.04 is shown in Figure 3a,b. The analyzed building is an office building. The building in which the tests were carried out is a building that meets the requirements of the building "with almost zero energy demand" (NZEB). Buildings of this type will be standard in European Union countries from 2021. NZEB buildings are characterized by high wall insulation and high tightness of the building envelope. The building is equipped with a heating and cooling system, the power of which has been calculated assuming that the system works in the summer and winter season. The system was not expected to work during the transition period. Most charts present round the clock measurements, but the detailed analysis focuses on working hours. The rooms analyzed are small office rooms with an area of about 37 m^2. The research assumed that each room is used by one person between 8.00 a.m. and 8.00 p.m.

The parameters of the sensors are presented in Table 2.

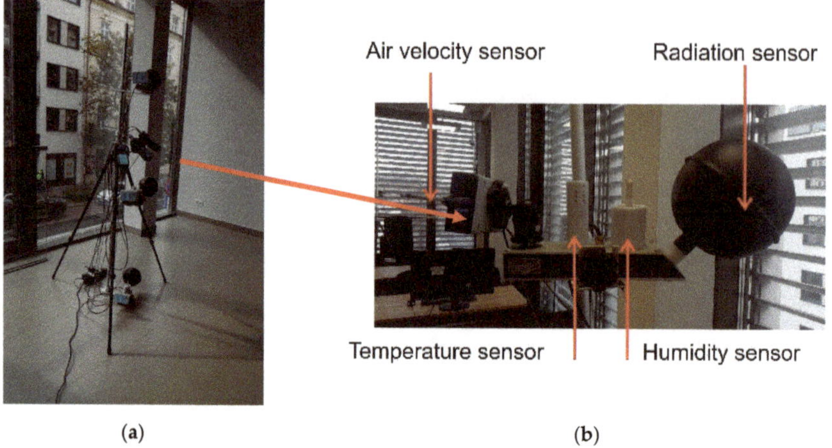

(**a**) (**b**)

Figure 3. (**a**) View of microclimate meter. (**b**) Description of sensors.

Table 2. Sensors' specifications.

Type of Sensor	Measurement Range	Scale	Accuracy
Temperature Sensors	−20 °C + 50 °C (wet thermometer 0 °C + 50 °C)	0.01 °C	±0.4 °C
Humidity Sensors	0–100%	0.1 RH (relative humidity)	±2% RH (relative humidity)
Air Velocity Sensors	0–5 m/s	0.01 m/s	for 0–1 m/s +/0.05 + 0.05 × Va m/s, for 1–5 m/s ± 5%

Figure 4 presents the arrangement of the test equipment in rooms P1.06 and P2.04. The microclimate meter is placed at the intended workstation.

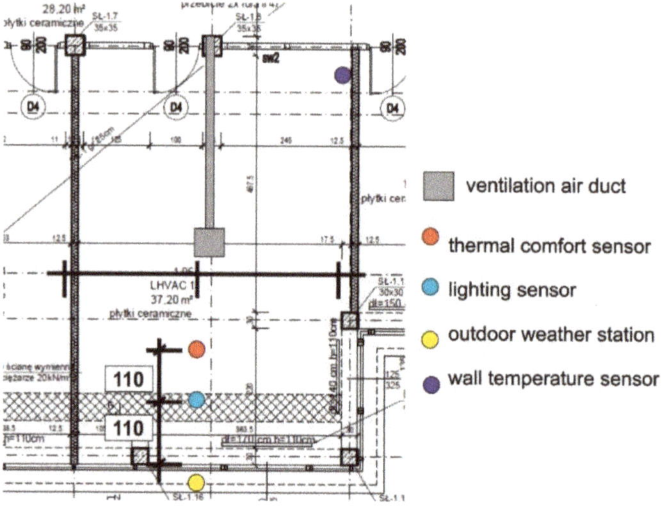

Figure 4. Arrangement of sensors in rooms P1.06 and P2.04.

The tests were carried out using measuring equipment that meets the requirements of the PN–EN 7726 [33] standard. The measuring equipment (Figure 3) is a microclimate meter.

The measured parameters included:

t_a ambient air temperature;

t_g temperature of blackened sphere (heat radiation meter)—the black sphere, according to the standards, should have a diameter of 15 cm;

t_{nw} natural wet-bulb temperature;

RH   relative air humidity;

V_a air flow rate.

The data were collected every 10 min.

The data from the sensors are given in Table 2.

On the basis of measurements, thermal comfort parameters PMV and PPD were calculated from formulas [9,34,35].

External environmental parameters were collected from weather stations located at the laboratory's south elevation (Figure 5). The weather station recorded the following data (Figure 6):

- Outdoor temperature T_e °C
- Relative humidity (outdoor) RH_e %
- Outdoor air velocity a_e m/s
- Total outdoor radiation intensity I_e W/m^2

Figure 5. Weather station on the southern façade.

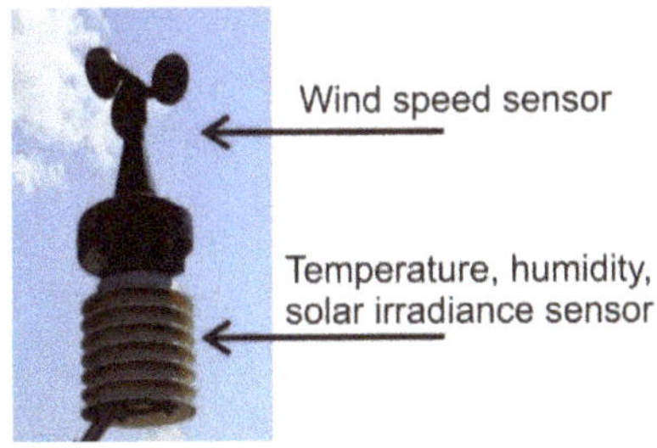

Figure 6. Description of sensors in the weather station.

The scope of measurement and accuracy of the sensors in the weather station are presented in Table 3.

Table 3. Data of weather station sensors.

No.	Location	Measured Parameter	Measurement Scope	Accuracy
1		Wind speed a_e m/s	0.00–50.00 m/s (possible max. = 60)	+/−1.0 m/s or +/−5%
2	Weather station on the southern façade	Relative humidity (outdoor) RH_e %	0 ÷ 90%	+/−3%
3		Relative humidity (outdoor) RH_e %	90 ÷ 100%	+/−4%
4		Outdoor temperature T_e °C	−40 ÷ 60 °C	+/−0.5 °C
5		Total solar radiation intensity i_e W/m^2	0 ÷ 1500 W/m^2 (possible max. = 2000)	+/−5%

3. Results

Sunny days with minor cloud cover dominated in the period selected for the analysis, i.e., from 27 March to 6 April 2017. As a result of minor cloud cover, the air temperature during the day was fairly high but low during the night. Figure 7 presents the waveform of locally measured values (weather station on the southern façade) of outdoor air temperature and total solar radiation intensity. In Figures 7–17 the hours of use of the rooms (8 a.m.–8 p.m.) are marked with a rectangle.

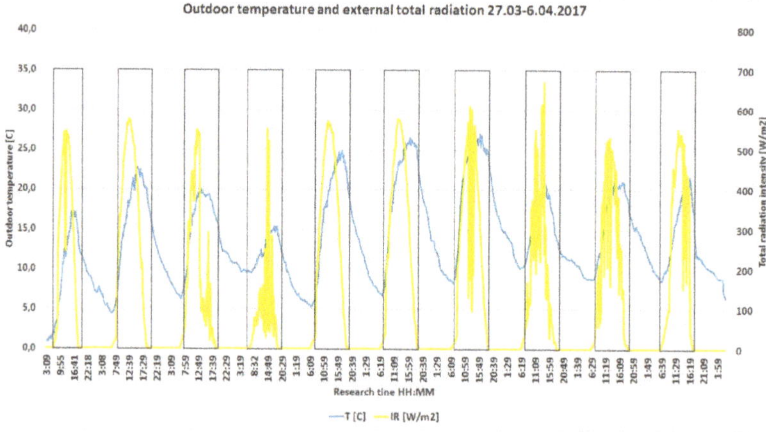

Figure 7. Waveform of outdoor air temperature and total solar radiation between 27 March and 6 April 2017. Measurements were carried out by means of a weather station on the southern façade.

No heating or cooling systems were operated in the test rooms, as was described in the "Materials and Methods" section. The rooms were not used in the reference period and, as such, no internal heat gains were generated. The only external façade of rooms P1.06 and P2.04 is glazed. Other walls were adjacent to the rooms, where there were constant temperature conditions (+20 °C).

3.1. Room P1.06

Figure 8 presents the wavelength of outdoor and indoor temperatures in room P1.06—with no venetian blinds installed.

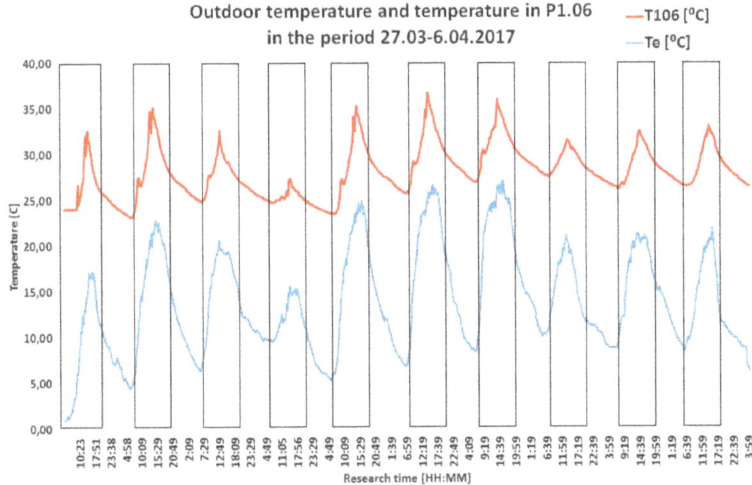

Figure 8. Waveform of outdoor and indoor temperatures in room P1.06—with no venetian blinds installed.

The duration of maximum and minimum air temperature values in room P1.06 corresponds to the wavelength of the maximum and minimum outdoor temperatures. This is shifted by about two hours against the maximum values of solar light intensity for the southern façade.

The ambient air temperature in the analysed room reaches very high values. From 23.25 °C at night, to 35.13 °C in the early afternoon. Despite low outdoor temperatures at night (decreasing to ca. 5 °C), intensive solar radiation during the day (max. value 672 W/m^2) contributes to such a temperature wavelength. Another factor which affects temperature distribution in the room is the continuous air supply at a constant temperature of 18°C and the stabilised temperature conditions in the adjacent rooms (P1.04, P1.07 and P1.10).

Figure 9 presents indoor temperature (T_i), mean radiation temperature (T_r) and operating temperature T_o in room P1.06. The T_r value was identified based on the recorded temperature measurements of a blackened sphere. Significant differences, which sometimes exceed 20%, can be observed.

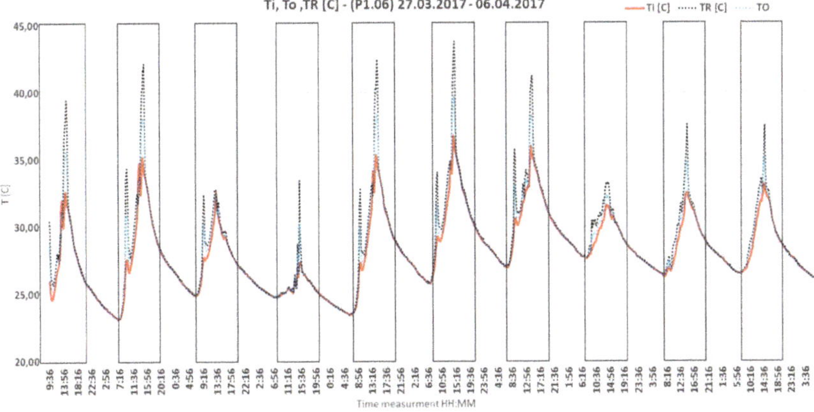

Figure 9. Ambient air temperature (T_i °C) and infrared radiation temperature (TR °C) in room P1.06.

Differences between T_i, T_R and, consequently, the operating temperature T_o are clearly noticeable during the hours of intensive solar radiation. Sample wavelengths of the temperature values for the selected day are presented in Figure 10. At high indoor temperature that day in room P1.06 (26.9 °C–35.6 °C), the operating temperature reached the maximum value of 38.3 °C.

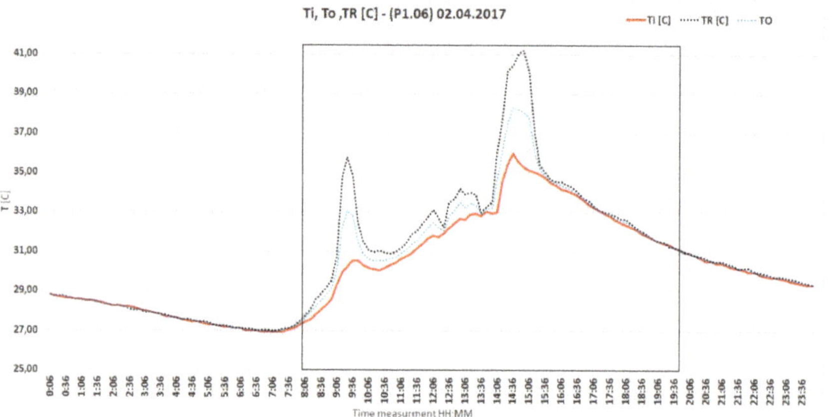

Figure 10. Values of ambient air temperature T_i, mean temperature of a blackened sphere TR and resultant operating temperature T_o (Room P1.06) on 2 April 2017.

3.2. Room P2.04

Room P2.04 is equipped with a system of external sun shading devices—venetian blinds—with panels that can be positioned at any angle. For the entire period of the experiment, the venetian blinds were positioned at an angle of 45°. Figure 11 shows the values of total solar radiation on the southern façade, indoor air temperature in rooms P1.05 and P2.04 and external temperature.

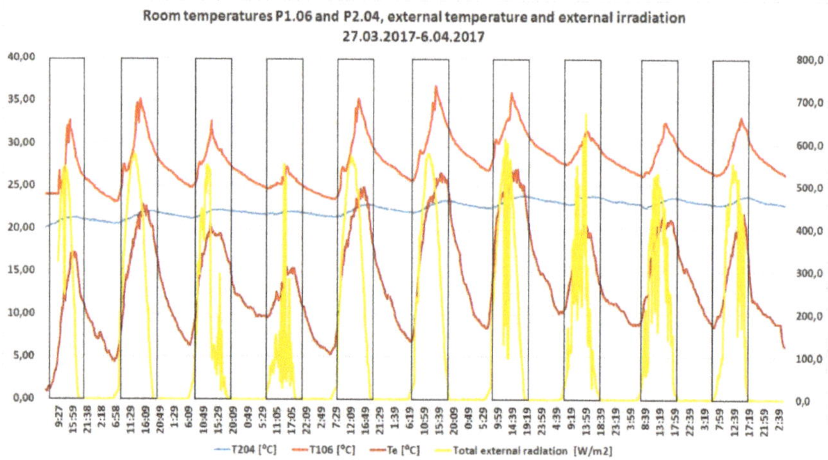

Figure 11. Wavelength of external radiation and indoor temperature in room P2.04—with external sun shading devices.

Figure 12 presents a wavelength of outdoor and ambient air temperatures in room P2.04.

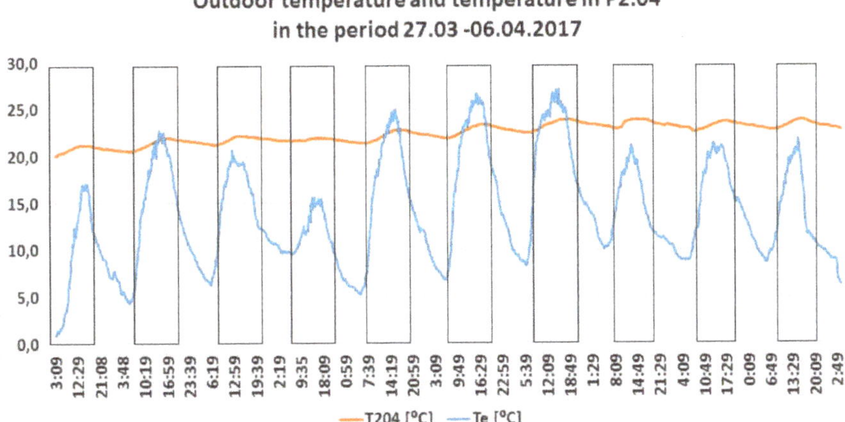

Figure 12. Wavelength of outdoor and indoor temperatures in room P2.04—with external sun-shading devices.

Ambient air temperature in room P2.04 was highly stabilised and ranged from 20.13 °C to 23.80 °C in the analysed period. This result is mainly caused by the significant reduction in solar gain owing to the system of external venetian blinds. The external panels were positioned at a 45° angle towards the windowpane. The nearly unchanged temperature values in room P2.04 were also related to the constant air supply temperature (18 °C) and the stabilised temperature conditions in the adjacent rooms (P2.03, P2.05 and P2.08). Solid reinforced concrete floors with a high heat capacity, limiting the room at the top and bottom, were another factor which alleviated temperature fluctuations.

The maximum temperature in room P2.04 was normally recorded in late afternoon hours. The time difference between the value of the greatest solar radiation intensity and the highest temperature in room P2.04 was up to six hours. This resulted from the position (inclination) of the panels of the external venetian blinds, which protected the room from direct solar radiation gain contrary to room P1.06, which was not equipped with a system of sun shading devices.

As a result of using external sun shading devices in room P2.04, the ambient air temperature distribution and the value of the mean radiation temperature were stable, which contributed to a favourable distribution of operating temperature (Figure 13).

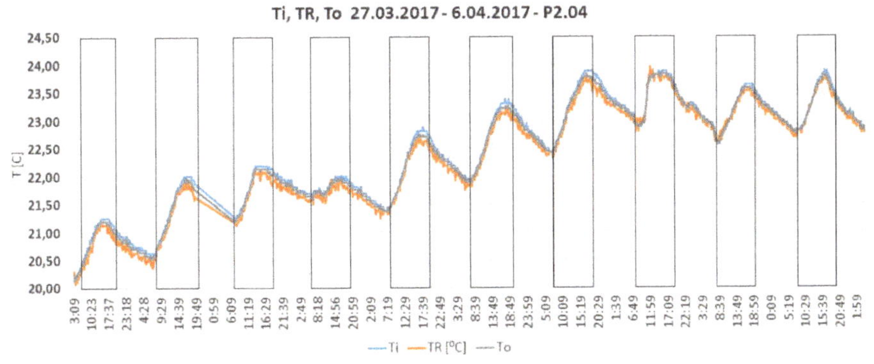

Figure 13. Wavelength of indoor air temperature changes in room P2.04 (T_i) and mean radiation temperature (TR).

In this case, contrary to room P1.06, the values of ambient air temperature T_i, radiation T_R and T_o were similar, even during intensive solar radiation hours. The recorded differences do not exceed 0.25 °C.

The identified mean radiation temperature value for the room with external venetian blinds is lower than the ambient air temperature recorded during that time. The opposite occurs in room P1.06 (Figures 9 and 10), where solar radiation penetration was strong. The mean surface radiation temperature reaches much higher values than ambient air temperature.

Figure 14 presents a wavelength of outdoor and ambient air temperatures in rooms P1.06 (with no venetian blinds) and P2.04 (with external shading devices).

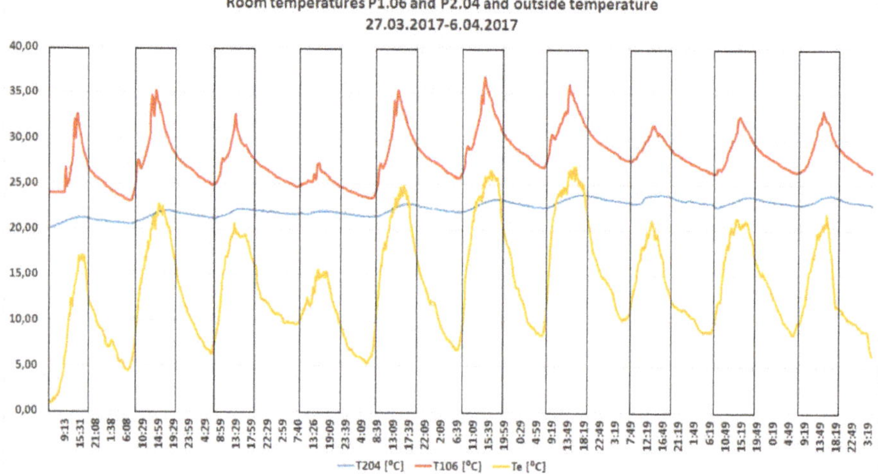

Figure 14. Wavelength of outdoor and ambient air temperatures in rooms P1.06 (with no venetian blinds) and P2.04 (with external shading devices). Visible data gaps are caused by technical problems with the measuring equipment. The sensors had to be submitted for technical inspection.

3.3. Comparison of Results

When identifying thermal comfort indices PMV and PPD, it was assumed for both rooms that their users worked in clothes characteristic for the winter/spring season. The thermal insulation values of the clothing, Clo, were adopted based on standard [9], which applies to people doing office work.

The Clo and Met values adopted for analysis are shown in Table 4.

Table 4. Clo and Met values adopted for thermal comfort analyses for the transition season between 27 March and 6 April 2017.

Room	Clothing Insulation Properties Clo	Metabolism Met
P1.06	1.0	1.21
P2.04	1.0	1.21

The results of analysis and measurements of PMV and PPD comfort indices for both rooms are presented in Figures 15 and 16.

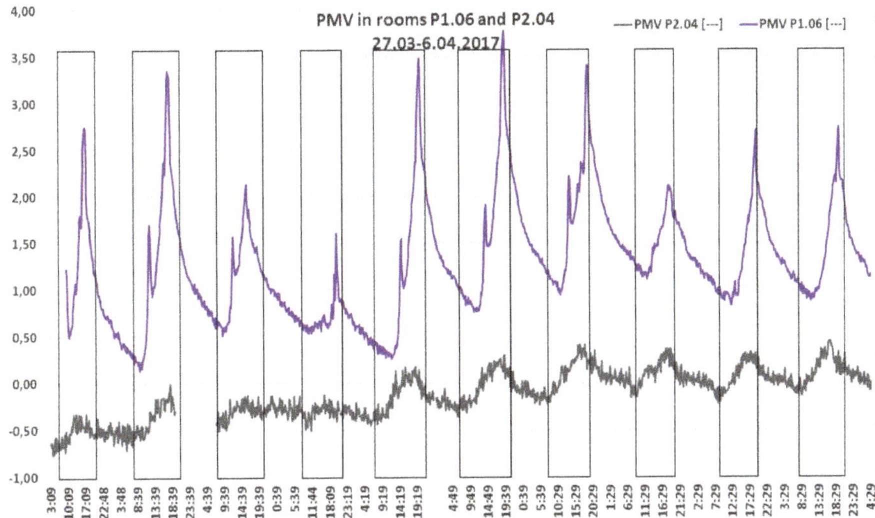

Figure 15. Values of predicted mean vote (PMV) index in rooms P1.06 and P2.04. Visible data gaps are caused by technical problems with the measuring equipment. The sensors had to be submitted for technical inspection.

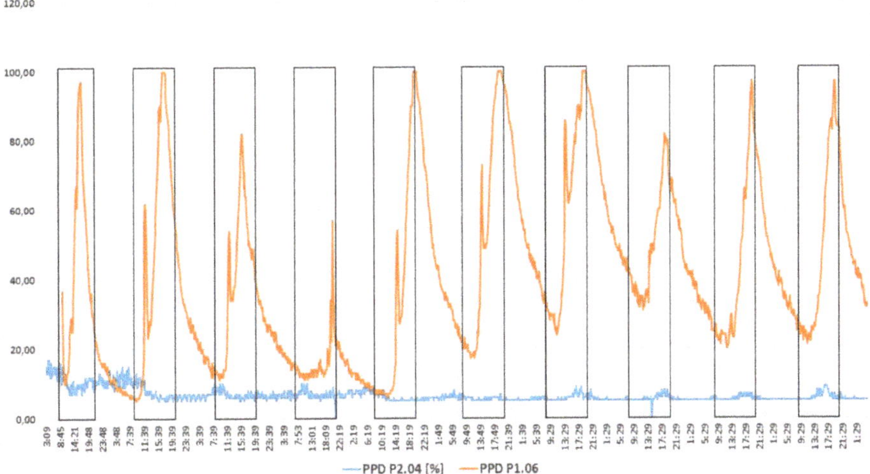

Figure 16. Values of predicted percentage of dissatisfied (PPD) index in rooms P1.06 and P2.04.

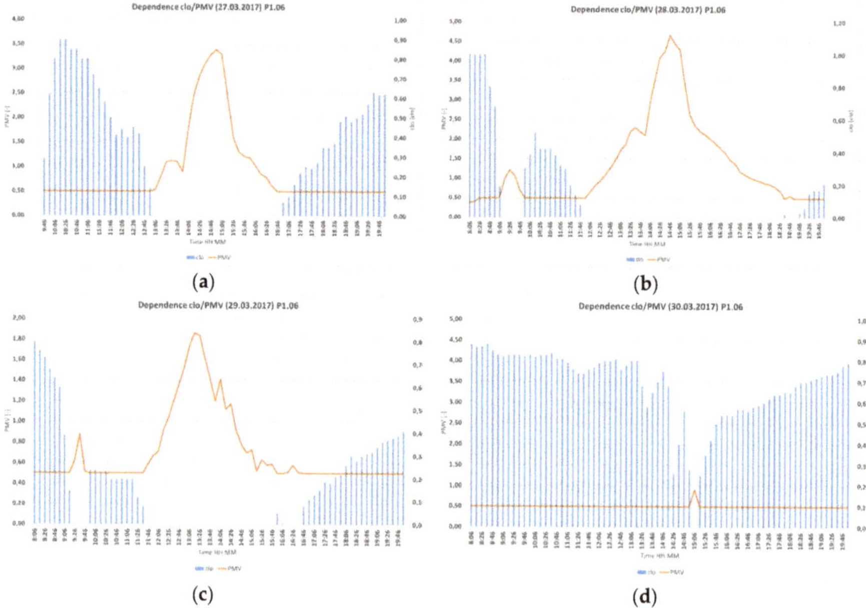

Figure 17. (**a**) PMV relationship (assumed value: +0.5 [-]) P1.06. (27.03.2017); (**b**) P1.06. (28.03.2017); (**c**) P1.06. (29.03.2017); (**d**) P1.06. (30.03.2017).

4. Discussion

The wavelengths of the values of the PMV and PPD indices are presented in Figures 15 and 16. They clearly suggest the high efficiency of the external shading devices. The values of the PMV index in the room with closed venetian blinds ranged from −0.75 to +0.17. The percentage of dissatisfied people amounted to 5–17%. This means that the conditions in the room both during the day and at night were nearly perfect. The number of thermal comfort hours in room P2.04 (external venetian blinds installed) in the analysed period was 23, with only nine hours between 8.00 a.m. and 10.00 p.m. These hours can be treated as the office hours, which means that users do not stay in the building beyond them. A large part of the energy consumed by office buildings is used for cooling. This energy can be significantly reduced owing to shading devices. Many researchers have demonstrated the efficiency of such measures through simulations [32,36]. A simulation study in southern Italy, presented in [37], showed the possibility of reducing energy consumption for cooling purposes in highly glazed office rooms with external roller blinds by nearly 50%, compared to rooms with no external roller blinds installed. In the room marked as P1.06 (with no external sun-shading devices), conditions regarded as comfortable lasted for only 16 h. For the rest of the time, temperatures in the range of 27 °C and 35 °C were unacceptable for the users. The maximum PMV value amounts to 3.48 at 100% of dissatisfied people (PPD = 100%). A number of researchers has studied the behaviour of users of rooms in order to improve ambient air quality, thermal comfort and building performance efficiency (performance improvement). The opening of windows is among such behaviours [38] and [39]. Studies of thermal comfort [40], presented by the team from Harbin, covered transition seasons, in addition to the winter heating season. The behaviours of people aimed at adjusting their thermal comfort were analysed for the spring season, also when heat distribution systems were not in operation. The users of office and residential premises improved their comfort by opening windows, changing the insulation characteristics of their clothing and by drinking more water. Windows were not opened in the rooms of the MLBE building analysed in the paper. The authors analysed the extent to which a change in the insulation characteristics of clothes (Clo = 1.0 assumed in the tests) may contribute to the thermal comfort experience of the users under the conditions that occurred in room P1.06. The Clo value at which the thermal comfort of users would become neutral was calculated. The results are presented in Figure 17a–d. For selected days (27–30 March), the figures present the Clo value (blue bars) at which the users of the room would have to reduce the insulation characteristics of their clothes to reach an acceptable level of thermal comfort experience. For the analysed office building, the PMV should range from −0.5 to +0.5.

Analysing the results presented for sunny days (Figure 17a–c), it can be concluded that despite highly uncomfortable (unfavourable) thermal conditions in the room, user behaviour can improve the comfort experienced. Such improvement is possible in the pre-noon and afternoon hours. At noon and in the early afternoon, no clothing reduction helped to reach comfortable conditions.

On a cloudy day (30 March) with lower solar radiation intensity, comfortable conditions can be reached by slightly changing the insulation characteristics of clothing (clo) to the values of 0.62–0.92 (Figure 17d). Only the calculations for the period between 14:00 and 16:00 revealed a significant decrease in the clothing insulation characteristics due to the momentary operation of the sun (Figure 11).

An analysis of the selected period between 27 March and 6 April 2017 revealed that when the users of the rooms with no external venetian blinds changed their clothing insulation characteristics, the number of discomfort hours reduced from 117 to 81 (during the hours of use of the rooms).

5. Conclusions

The building in which the tests were carried out meets the requirements of the building "with almost zero energy demand" (NZEB). Buildings of this type will be standard in European Union countries from 2021. NZEB buildings are characterized by high wall insulation and high tightness of the building envelope. The building is equipped with a heating and cooling system, the power of which has been calculated assuming that the system works in the summer and winter season. The system was not expected to work during the transition period. Studies have shown that in the case of large glazed facades, overheating also occurs during the transition period, especially during cloudless sunny days. Such a building's response in the analyzed time may contribute to the need to cool the rooms, and thus to increase operating costs.

Highly glazed façades in office rooms may generate significant internal thermal gains even in the transition season (spring). External venetian blinds used in the reference room helped to maintain comfortable thermal conditions ($-0.5 <$ PMV $< +0.5$). This means that the implementation of such systems helps to reduce or even eliminate energy consumption for cooling purposes.

The aforementioned thesis is confirmed by the thermal comfort conditions observed during the test in the room with no venetian blinds. Despite the same geometry, orientation, use and equivalent schedule of ventilation equipment operation, the thermal comfort conditions obtained ($0.20 <$ PMV < 3.70) varied significantly from those in the room with venetian blinds.

Another conclusion drawn based on the experiment conducted concerns the possibility of the users improving their own thermal comfort experience. Replacing clothes recommended for the heating season (Clo = 1.0) with summer office clothes ($0.25 <$ Clo < 1.0) helped to reduce the number of thermal discomfort hours of employees by 30%. Such actions can reduce the number of operating hours of the cooling system from 12 h to even less than 4 h a day. This was a case study carried out on chilly but sunny days of the transition season.

Author Contributions: Conceptualization, M.F.-C., M.F.; methodology, M.F.-C., K.N.; software, M.F.-C., K.N.; validation, M.F.-C., K.N.; formal analysis, M.F.-C., K.N.; investigation, M.F.-C., K.N.; resources, M.F.-C., M.F., K.N.; data curation, M.F.-C., K.N.; writing-original draft preparation, M.F.-C., K.N.; writing-review and editing, M.F.-C., K.N.; visualization, M.F.-C., K.N.; supervision, M.F.-C., K.N.; project administration, M.F.-C., K.N.; funding acquisition, M.F.-C., M.F. All authors have read and agreed to the published version of the manuscript.

Funding: This research received no external funding.

Acknowledgments: The authors thanks Aleksander Panek, Michał Piasecki and Agnieszka Lechowska for substantive support.

Conflicts of Interest: The authors declare no conflict of interest.

References

1. Dyrective 2002/91/UE; DIRECTIVE 2002/91/EC of the European Parliament and of the Council of 16 December 2002 on the Energy Performance of Buildings. 2002. Available online: https://eur-lex.europa.eu/legal-content/PL/TXT/?uri=celex%3A32002L0091 (accessed on 19 December 2019).
2. Directive 2010/31/EU of the European Parliament and of the Council of 19 May 2010 on the Energy Performance of Buildings. Available online: https://eur-lex.europa.eu/legal-content/EN/TXT/?uri=CELEX%3A32010L0031 (accessed on 19 December 2019).
3. OBWIESZCZENIE MINISTRA INFRASTRUKTURY I RO ZWOJU1) z dnia 17 lipca 2015 r. w Sprawie ogłoszenia Jednolitego Tekstu Rozporządzenia Ministra Infrastruktury w Sprawie Warunków Technicznych, Jakim Powinny Odpowiadać Budynki i ich Usytuowanie. Available online: http://prawo.sejm.gov.pl/isap.nsf/DocDetails.xsp?id=WDU20150001422 (accessed on 19 December 2019).
4. Wymagania Dla Budynków Pasywych. Available online: http://www.pibp.pl/ (accessed on 15 September 2019).
5. Bao, X.; Tian, Y.; Yuan, L.; Cui, H.; Tang, W.; Fung, W.H.; Qi, H. Development of high performance PCM cement composites for passive solar buildings. *Energy Build.* **2019**, *194*, 33–45. [CrossRef]
6. Kwasnowski, P.; Fedorczak-Cisak, M.; Knap, K. Problems of technology of energy-saving buildings and their impact on energy efficiency in buildings. In Proceedings of the 2017 IOP Conference Series: Materials Science and Engineering, Volume 245: World Multidisciplinary Civil Engineering-Architecture-Urban Planning Symposium—WMCAUS, Prague, Czech Republic, 12–16 June 2017; Available online: https://iopscience.iop.org/article/10.1088/1757-899X/245/7/072043/pdf (accessed on 19 December 2019). [CrossRef]
7. Fanger, P.O. *Thermal Comfort: Analysis and Applications in Environmental Engineering*; McGraw-Hill: New York, NY, USA, 1972; ISBN 9780070199156.
8. Fanger, P.O. *Analysis and Applications in Environmental Engineering*; McGraw-Hill Book Company: New York, NY, USA, 1970; ISBN 0070199159.
9. ISO Standard 7730. *Ergonomics of the Thermal Environment-Analytical Determination and Interpretation of Thermal Comfort Using Calculation of the PMV and PPD Indices and Local Thermal Comfort Criteria*; International Organization for Standardizati: Geneva, Switzerland, 2005.
10. PN-EN 15251:2012—Wersja Polska. Available online: http://sklep.pkn.pl/pn-en-15251-2012p.html (accessed on 19 September 2019).
11. PN-EN 16798-1:2019-06—Wersja Angielska. Available online: http://sklep.pkn.pl/pn-en-16798-1-2019-06e.html (accessed on 19 September 2019).
12. Yao, R.; Li, B.; Liu, J. A theoretical adaptive model of thermal comfort—Adaptive predicted mean vote (aPMV). *Build. Environ.* **2009**, *44*, 2089–2096. [CrossRef]
13. Van Hoof, J.; Mazej, M.; Hensen, J.L.M. Thermal comfort: Research and practice. *Front. Biosci.* **2010**, *15*, 765–788. [CrossRef]
14. Mıhlayanlar, E.; Öztuna, S.; Büyükakın, K. Investigation of thermal comfort conditions in higher education facilities: A case study for engineering faculty in edirne. *TEM J.* **2017**, *6*, 71–79.
15. Yang, L.; Yan, H.; Lam, J.C. Thermal comfort and building energy consumption implications—A review. *Appl. Energy* **2014**, *115*, 164–173. [CrossRef]
16. Lai, A.C.K.; Mui, K.W.; Wong, L.T.; Law, L.Y. An evaluation model for indoor environmental quality (IEQ) acceptance in residential buildings. *Energy Build.* **2009**, *41*, 930–936. [CrossRef]
17. Piasecki, M.; Kostyrko, K.; Pykacz, S. Indoor environmental quality assessment: Part 1: Choice of the indoor environmental quality sub-component models. *J. Build. Phys.* **2017**, *41*, 264–289. [CrossRef]
18. Piasecki, M.; Barbara Kostyrko, K. Indoor environmental quality assessment, part 2: Model reliability analysis. *J. Build. Phys.* **2018**, *42*, 288–315. [CrossRef]
19. Nimlyat, P.S. Indoor environmental quality performance and occupants' satisfaction [IEQPOS] as assessment criteria for green healthcare building rating. *Build. Environ.* **2018**, *144*, 598–610. [CrossRef]

20. Damiati, S.A.; Zaki, S.A.; Wonorahardjo, S.; Ali, M.S.M.; Rijal, H.B. Thermal comfort survey in office buildings in Bandung, Indonesia. In Proceedings of the International Joint Conference SENVAR-iNTA-AVAN, Bahru, Malaysia, 24–26 November 2015; pp. 53–64.

21. Zmeureanu, R.; Bessoudo, M.; Tzempelikos, A.; Athienitis, A. The impact of shading on thermal comfort conditions in perimeter zones with glass facades. In Proceedings of the 2nd PALENC Conference and 28th AIVC Conference on Building Low Energy Cooling and Advanced Ventilation Technologies in the 21st Century, Crete, Greece, 27–29 September 2007; pp. 1072–1077.

22. Palmer, J.; Bennetts, H.; Pullen, S.; Zuo, J.; Ma, T.; Chileshe, N. Adaptation of Australian houses and households to future heat waves. In Proceedings of the 7th Australasian Housing Researchers' Conference, AHRC 2013, Fremantle, Australia, 6–8 February 2013.

23. Kisilewicz, T. Passive control of indoor climate conditions in low energy buildings. *Energy Procedia* **2015**, *78*, 49–54. [CrossRef]

24. Kajtár, L.; Erdősi, I.; Bakó-Bíró, Z. Thermal and air quality comfort of office buildings based on new principles of dimensioning in hungary. *Periodica Polytechnica Ser. Mech. Eng.* **2000**, *44*, 265–274.

25. Kavgic, M.; Mumovic, D.; Stevanovic, Z.; Young, A. Analysis of thermal comfort and indoor air quality in a mechanically ventilated theatre. *Energy Build.* **2008**, *40*, 1334–1343. [CrossRef]

26. Kisilewicz, T. *Wpływ izolacyjnych, Dynamicznych i Spektralnych Właściwości Przegród na Bilans Cieplny Budynków Energooszczędnych*; Cracow University of Technology Publisher: Kraków, Poland, 2008; ISSN 0860-097X.

27. Alwetaishi, M. Impact of glazing to wall ratio in various climatic regions: A case study. *J. King Saud Univ. Eng. Sci.* **2019**, *31*, 6–18. [CrossRef]

28. Lin, Y.; Zhou, S.; Yang, W.; Li, C.Q. Design optimization considering variable thermal mass, insulation, absorptance of solar radiation, and glazing ratio using a prediction model and genetic algorithm. *Sustainability* **2018**, *10*, 336. [CrossRef]

29. Palmero-Marrero, A.I.; Oliveira, A.C. Effect of louver shading devices on building energy requirements. *Appl. Energy* **2010**, *87*, 2040–2049. [CrossRef]

30. Bellia, L.; Marino, C.; Minichiello, F.; Pedace, A. An overview on solar shading systems for buildings. *Energy Procedia* **2014**, *62*, 309–317. [CrossRef]

31. Atzeri, A.; Cappelletti, F.; Gasparella, A. Internal versus external shading devices performance in office buildings. *Energy Procedia* **2014**, *45*, 463–472. [CrossRef]

32. Nowak, K.; Nowak-Dzieszko, K.; Rojewska-Warchal, M. Thermal comfort of the rooms in the designing of commercial buildings. In *Research and Applications in Structural Engineering, Mechanics and Computation—Proceedings of the 5th International Conference on Structural Engineering, Mechanics and Computation (SEMC) 2013, Cape Town, South Africa, 2–4 September 2013*; CRC Press Taylor&Francis Group: Boca Raton, FL, USA, 2013; pp. 1819–1824.

33. PN-EN ISO 7726:2002 Ergonomics of the Thermal Environment—Instruments for Measuring Physical Quantities (ISO 7726:1998). Available online: https://www.sis.se/api/document/preview/615884/ (accessed on 19 December 2019).

34. DIN EN 15251—European Standards. Indoor Environmental Input Parameters for Design and Assessment of Energy Performance of Buildings Addressing Indoor Air Quality, Thermal Environment, Lighting and Acoustics. Available online: https://www.en-standard.eu/din-en-15251-indoor-environmental-input-parameters-for-design-and-assessment-of-energy-performance-of-buildings-addressing-indoor-air-quality-thermal-environment-lighting-and-acoustics/ (accessed on 19 December 2019).

35. PN-EN 16798-1:2019-06 Energy Performance of Buildings—Ventilation for Buildings—Part 1: Indoor Environmental Input Parameters for Design and Assessment of Energy Performance of Buildings Addressing Indoor Air Quality, Thermal Environment, Lighting a. 2019. Available online: https://shop.bsigroup.com/ProductDetail/?pid=000000000030297474 (accessed on 19 December 2019).

36. Al-Tamimi, N.A.; Fadzil, S.F.S. The potential of shading devices for temperature reduction in high-rise residential buildings in the tropics. *Procedia Eng.* **2011**, *21*, 273–282. [CrossRef]

37. Evola, G.; Gullo, F.; Marletta, L. The role of shading devices to improve thermal and visual comfort in existing glazed buildings. *Procedia Eng.* **2017**, *134*, 346–355. [CrossRef]

38. Li, N.; Li, J.; Fan, R.; Jia, H. Probability of occupant operation of windows during transition seasons in office buildings. *Renew. Energy* **2015**, *73*, 84–91. [CrossRef]

39. Tanner, R.A.; Henze, G.P. Stochastic control optimization for a mixed mode building considering occupant window opening behaviour. *J. Build. Perform. Simul.* **2014**, *7*, 427–444. [CrossRef]
40. Ning, H.; Wang, Z.; Ren, J.; Ji, Y. Thermal comfort and thermal adaptation between residential and office buildings in severe cold area of China. *Procedia Eng.* **2015**, *121*, 365–373. [CrossRef]

Review

Future Design Approaches for Energy Poverty: Users Profiling and Services for No-Vulnerable Condition

Andrea Boeri *, Valentina Gianfrate, Saveria Olga Murielle Boulanger and Martina Massari

Architecture Department, University of Bologna, 40136 Bologna, Italy; valentina.gianfrate@unibo.it (V.G.); saveria.boulanger@unibo.it (S.O.M.B.); m.massari@unibo.it (M.M)
* Correspondence: andrea.boeri@unibo.it

Received: 30 January 2020; Accepted: 21 April 2020; Published: 24 April 2020

Abstract: Analyzing data from the Energy Poverty Observatory in Europe, it emerges that more than 50 million households in the EU live in energy poverty (people that cannot heat their homes during winter; cannot make their homes comfortable during the summer; pay their energy bills late). Research studies realized in the last 20 years highlight that making energy demand efficient and effective is the more significant and socially important the more it is able to involve users who are unable to sustain energy demand. The evolution of the research sees a narrowing of the field of investigation by focusing on the user dimension of energy poverty, stressing the role of citizens not only as consumer but also as producers of solutions to tackle energy poverty, real energy communities of agents. The paper aims to provide a systematic literature review highlighting the major findings of the topic, investigating the relationship between spatial and social issues, and looking at the state of energy poverty by addressing the profiling of users and consequently of services useful to overcome their current vulnerable condition. The paper is structured in two core sections. The first one gives the results of a systematic literature review on the energy/fuel poverty topic, the second one deepens the role of communities and individuals need, crucial in defining new design approaches for supportive solutions to tackle energy poverty.

Keywords: energy poverty; vulnerable users; energy communities; energy poverty metrics

---

1. Introduction

Energy poverty is gaining attention from European [1,2] global [3,4] policies and research paths [5,6] for more than a decade. The concept identifies a situation where a family or individual does not achieve an adequate level of essential energy services often due to a cascade of conditions, including low income and high energy expenditure caused by poor energy efficiency of housing [7].

It is a complex and often fragmented problem crossing several disciplines [8,9] affecting from 50 to 150 million people who are unable to pay for the primary energy services—such as heating, cooling, lighting, travel, and electricity—that are needed to ensure a decent standard of living [10].

Based on pioneering studies [11,12], a household could be affected by energy poverty due to low income, high energy costs, and energy inefficient dwellings. In concrete terms, this means that vulnerable citizens [13] do not have access to energy services or that the use of these energy services compromises their ability to access other basic services. The Covenant of Mayors 2018 highlights that energy poverty can have serious consequences for health, well-being, social inclusion, and quality of life [14]. According to the report conducted by the International Energy Agency [15], energy poverty includes not only lack of access to modern energy services, but also the reliability of these services and concerns in the affordability of access. Households affected by energy poverty, experience inadequate levels of some essential energy services such as lighting, heating/cooling, use of household appliances, transport, and much more.

Data from the Energy Poverty Observatory show that, in Europe, more than 50 million households alone live in energy poverty conditions, and in particular:

- 57 million people are unable to heat their homes during winter;
- 104 million people are unable to access a minimum level of comfort during summer;
- 52 million people pay their energy bills late.

In Italy, the various existing indicators estimate that Italian households in energy poverty are between 2.2 and 4.3 million. About 9.4 million individuals [16–18] can be considered energy poor and vulnerable, as they are unable to sustain energy costs to maintain an adequate level of comfort. For this reason, energy poverty has been seriously taken into account by several policy areas and agendas concerning social, economic, political, environmental, health [19], and climate. In particular, the European Commission's "Clean Energy for All European Citizens" [20] legislative package is providing measures to strengthen the position of the consumers, by providing rules that will give them more flexibility while also protecting them. The package focuses also on self-determination of the consumer, who is allowed to take its own decisions on "how to produce, store, sell or share energy" [20].

The European Commission (EC) has long been committed in tackling the problem, reinforcing its engagement with the establishment of the EU Energy Poverty Observatory in January 2018—to measure, monitor, and share knowledge and good practices to manage energy poverty—and the growing number of initiatives funded and promoted on the subject. The Covenant of Mayors for 2030, in addition to "taking action to mitigate climate change and adapt to its inevitable effects, requires signatories to commit to providing access to secure, sustainable and affordable energy for all. Covenant signatories can improve the quality of life of their citizens and create a fairer and more inclusive society through the reduction of energy poverty" [14].

The challenge of energy poverty has also been included in the United Nations' Agenda 2030 as one of the actions foreseen in Sustainable Development Objective 7 "Ensuring access to affordable, reliable, sustainable, and modern energy systems for all". These are just the last global orientation tackling a phenomenon that is emerging as a consequence of long-lasting macro-level [21] trends, such as the economic crisis, the obsolescence of the building stocks, the lack of qualitative data and knowledge about fuel poverty penetration, the heterogeneity of the phenomenon.

In 2009, EU GDP fell by 4.2% compared to 2008 and there was a sharp increase in unemployment. The economic and financial crisis that started in 2007 occurred against a background of falling wages for European workers. In addition, there was an increase in domestic energy prices, due to the EU's dependence on imported sources. From the end-user perspective, evidence suggest that the price of domestic energy in the EU has steadily increased since the mid-1990s, gradually reducing the purchasing power of households [22]. In the case of Italy for example, the increase in prices, while consumption was substantially stable, has led to an increase in energy expenditure, whose incidence on the total rose from 4.7% in 2007 to 5.1% in 2017. This created a worrying prospect for the most vulnerable households.

Furthermore, most Southern Europe buildings were built in the last century without (or with few) wall insulation, leading to inadequate winter performances. Not to mention the general low thermal transmittance values of surfaces, influencing thermal lag and, consequently, building summer performances. In fact, as assessed by researchers [23,24], fuel poverty is strongly linked with the thermal characteristics of buildings and fuel poor people are not only income poor. Thus, the topic is complex as it involves not only the necessity to create and boost national or local policies, but also the need to find methods for target identification, data sharing, and profiled services.

Energy transition and the resulting emergence of energy poverty take place against the background of marked differences between EU Member States. The geographical-spatial component is becoming increasingly important [25,26], together with the political-programmatic characterization of each state, and within them. For example, Italy is reducing its national energy-clean measures due to recent austerity policies but also to its political instability; this leaves room for the individual regions or even municipality to take over and produce operative but fragmented policies; Germany, on the other hand, is proposing a transformation of the energy sector with large-scale measures of energy efficiency and renewable energy solutions.

A growing amount of studies [13,27–29] highlight that making energy demand more efficient is all the more significant and socially important the more it is able to involve users who are unable to sustain energy demand. Therefore, contemporary research paths and European policy documents are narrowing the field of investigation by focusing on the user dimension of energy poverty, stressing the role of citizens not only as consumer, but also as producers of solutions to tackle energy poverty and real energy communities of agents.

Following these premises, the paper aims to read the concept in a European dimension not only by providing a five-years systematic literature review highlighting the most recent findings of the topic but also providing a qualitative analysis related to methods of measurement, mitigation measures, and users and vulnerabilities. In particular, the decision to focus mainly on the last five years was led by the intention to detect the most innovative and recent approaches to fuel poverty and to discover if the scientific debate is proceeding toward the inclusion of profiling studies.

In fact, the point of view of the relationship between the state of buildings and the state of poverty is investigated by addressing one of the determinants in the resolution of energy poverty: the profiling of users and consequently of services useful to overcome their current condition of poverty.

In order to answer to those research questions, the paper is structured in two core sections. The first one gives the results of a systematic literature review on the energy/fuel poverty topic having the objective to understand the main and more recent lines of development in the last five years; the second one performs a research aiming to deepens the role of communities, of vulnerabilities, and the potentialities of profiling services to address the problem. Finally, the conclusion recalls the main findings of the paper, assess some field limitations of the papers, and draft future research paths.

2. Research Methodology

The research underlying this paper is twofold, and it follows the structure of the paper. The first part of the paper follows in fact the methodology of the systematic literature review, implying the selection of a database, a timeframe, and specific queries, with the objective to analyze the most recent lines of development on the fuel and energy poverty topic.

The second part follows the methodology of a literature review of papers appeared in the most important international journal, without covering a specific timeframe and including both open access and not-open access papers. This review included the concepts of energy poverty measures, users and vulnerabilities and measures of mitigation. This second part of the research has the objective to give a more qualitative analysis of some of the main interesting aspects of the topic.

The systematic literature review has been performed in December 2019–January 2020 by using the Web of Science (WOS) database and by selecting open access and papers accessible from university libraries and online databases [30]. As a limitation to the field of investigation, the literature review has been done only considering papers written in English, as it is commonly recognized as the scientific international language and belonging to the European context. The results of this systematic review are showed in paragraph 3.

In the repositories, we searched works of the last five years (2015–2020) and we entered the following queries:

- "energy citizenship" on WOS, with a research in the entire paper text, 11 results
- "energy poverty", on WOS, entire paper text, 129 results
- "fuel poverty", on WOS, entire paper text, 150 results
- "energy vulnerability", on WOS, entire paper text, 32 results
- "energy poverty" AND "participation", on WOS, 12 results

- "fuel poverty" AND "participation", on WOS, 13 results
- "energy vulnerability" AND "participation", on WOS, entire paper text, 1 result.

The totality of the previous papers has been browsed and filtered following some rules as follows:

- unique entries (duplicate papers have been deleted);
- thematic relevance (for example papers focusing on media analysis such as TV channels and newspaper, or papers on transportation poverty, or on technical solutions not specifically linked with energy or fuel poverty were not considered);
- geographical relevance (only European fuel and energy poverty has been taken into account).

From these filters, 118 paper remained and were analyzed in depth. The analysis covered the following aspects:

- diffusion of the research and the related main themes in the last five years;
- thematic subdivision of papers;
- geographical distribution of case studies;
- identification of research red lines.

3. Analysis of Articles on European Energy Poverty: A Systematic Literature Review of the Most Recent Approaches

As introduced, energy and fuel poverty is a growing theme for European researchers. In the past years, in particular, the topic sees an increase in interest from several research fields. Not only the social science is interested in the topic but also architecture, medicine, and policies studies. On the one hand, in fact, the problem is more and more highlighted by policy-makers and professionals, and, on the other, more knowledge is produced, allowing more people searching innovative solutions.

On the semantic point of view, it is possible to observe how the two locutions of "fuel poverty" and "energy poverty" are generally considered as synonyms. Both, in fact, tend to include water, electricity, gas, and other fuels. A third locution appears frequently in the search: "energy justice" which can be considered a more general topic, not only working on technical solutions, but also including socio-political aspects.

On the chronological point of view, the systematic literature review (Figure 1) shows that almost the 35% of the total number of analyzed papers was written in 2019. Even with this predominance, the topic has been quite well investigated homogeneously in the last five years, with a small increase in 2017. An additional query on the WOS database about the keyword "energy poverty"—starting from 1995—showed how the topic has increased year by year since 1999, with a boost starting from 2010 (Figure 2), when the awareness on climate change and the role of building retrofitting in improving households conditions increased.

It is during these years that one of the first policies has been developed in the UK. In particular, the first UK policy is relevant as it sets a threshold below which a household can be considered fuel poor according to Boardman studies [31] (to spend more than 10% of the household income on energy in the home). Additionally, as described by Dubois and Meier [21], in 2009, a European project (EPEE) analyzed that, in the EU, fuel poverty involved between 50 and 125 millions of people and they put the attention to the connection between people vulnerability (especially in relation with elderly, disabled people and single-parent families) and the quality of buildings, highlighting how fuel poverty is more likely to happen in cold damp properties with insufficient heating system and insulation.

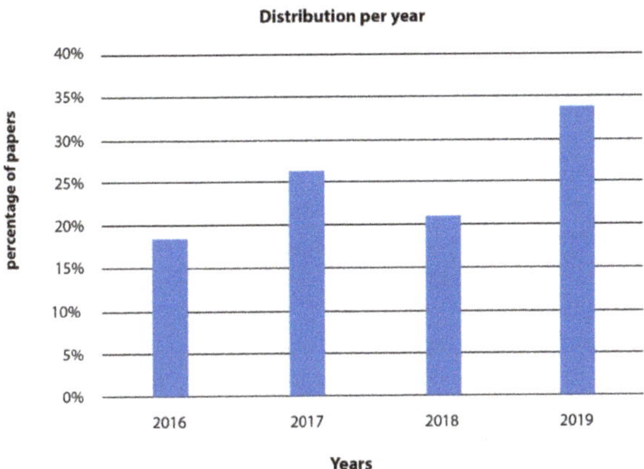

Figure 1. The image shows the distribution of the papers in the last few years.

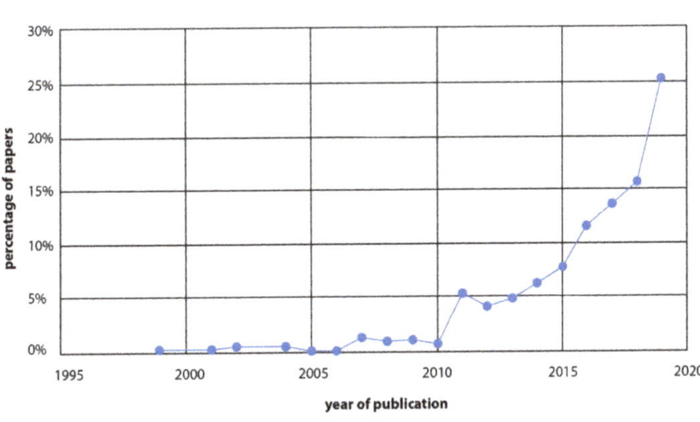

Figure 2. The image shows the growth of the number of papers from 2000 to 2019.

Considering the last five years, the selected papers can be clustered according to major recurring themes. In fact, 10 major categories have been found, as follows:

- welfare;
- theory (including also papers about semantic and definition);
- technical solutions (papers focusing on solving fuel/energy poverty on the technical point of view);
- stakeholders (papers focusing on the role of urban actors in mitigating fuel/energy poverty);
- profiling (papers focusing on specific targets);
- policies;
- participation (papers focusing on the role of participation and citizen engagement as a way to mitigate fuel/energy poverty);
- methods (papers proposing new methodologies to analyze the topic, also including indexes and indicators);
- health (papers specifically targeted to analyze the effect of fuel/energy poverty on health);
- behavioral studies (papers focusing on the role of people behaviors).

Before going deep into the identification of the main trends per each theme, it is important to notice that the majority of the selected paper apply their research and analysis on specific case studies. In fact, only a few papers, mainly related to the theory and method category, are untied from a specific case study. Furthermore, 3.58% of papers are related to a specific application. This finding is interesting as it shows how the topic is strictly linked with local contexts. As, in fact, argued by Dubois, "fuel poverty is strongly linked to the characteristics of housing […] indeed fuel poverty often results from a combination of low incomes and high energy needs" ([23], p. 107).

Among 58% of research applied to case studies, the high majority (51%) is referred to UK case studies, as visible in Figure 3. In fact, in Europe, the United Kingdom has been one of the first countries implementing specific national policies to solve it [2]. Thus, despite most part of the research seeming to come from the UK, several countries are strengthening their effort in addressing the topic, according to their individual policy framework. Secondly, in Spain (19%), there is a growing attention for the topic, while the rest of applications are mostly equally spread across Europe, with few highlights in Greece, Hungary, Ireland, Poland, and Portugal. Excluding the UK, most applications are concentrated on Southern Europe (39% of papers). This might be related to the effects of the economic crisis hitting harder in Southern Europe countries, provoking cascade negative effects on the construction industry—in charge of the management and improvement of the building's performances—and on the economic possibilities of the most fragile sector of the population. Furthermore, most Southern European buildings were built in the last century without (or with few) wall insulation, leading to inadequate winter performances. Not to mention the general low thermal transmittance values of surfaces, influencing thermal lag and, consequently, building summer performances In fact, data provided by the European Energy Poverty Observatory, in 2018, about the share of population not able to keep their home adequately warm, confirm how the worst situations are mainly concentrated in Southern and Eastern Europe with Bulgaria recording 33.7% of population in this situation, followed by Lithuania with the 27.9%, Greece (22.7%), Cyprus (21.9%), Italy 14.0%. In addition, data about arrears on utility bills seem to confirm this evidence: Spain 35.8%, Bulgaria 30.1%, Croatia 17.5%, Romania 14.4%, and Slovenia 12.5%. Even if these indicators are not sufficient to completely show the fuel poverty situation in Europe [32], they still seem interesting as they confirm the presence of emergencies in those regions and thus the necessity of increasing a knowledge and service creation effort for mitigating it.

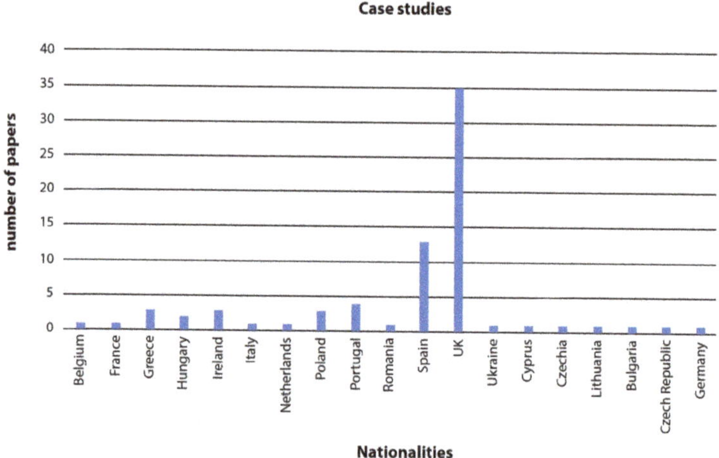

Figure 3. The graphic in the image represents the geographic distribution of the papers analyzed.

Assuming that a significant part of papers relies on local analysis, the thematic point of view is also interesting for the objectives of this research. In fact, given the 10 categories identified, it is interesting to note that most of them are spread in three major topics: methods and indicators (26%), theory (18%), and policies (16%) (Figure 4). This fact shows that the European debate about the topic seems to still be concentrated on answering research questions such as: "how can we measure the phenomenon?", "what is the extent of it?", "how can the fuel/energy poverty be defined?", "what does energy/fuel poverty mean?", "how can we increase our knowledge about it?". In particular, the identification of methods and indicators seems to be the most explored theme.

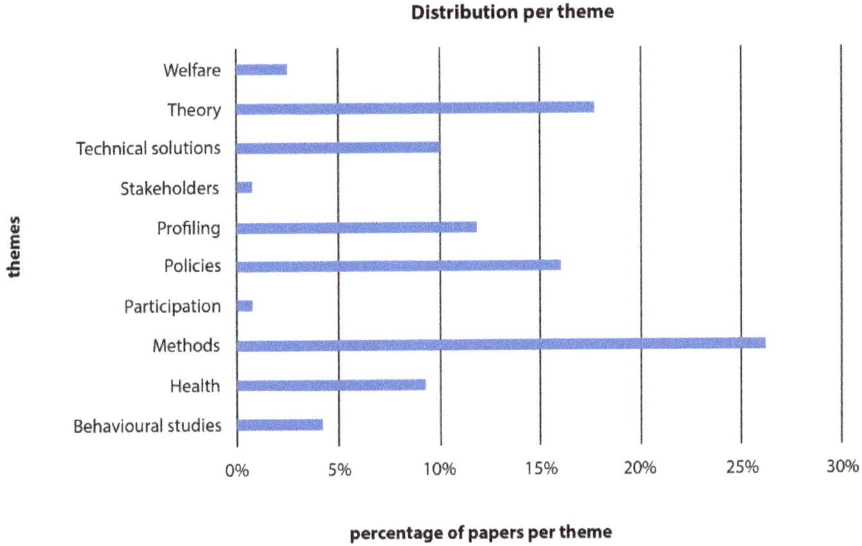

Figure 4. The graphic in the image represents the thematic distribution of the papers analyzed.

As described in several research works [33–35], the most used methodology for understanding the extension of the topic and for analyzing its characteristics in a specific context is to combine quantitative data from statistics (local, national, and European) with qualitative data. If the first category is more easy to find, the second one entails a deeper and closer relation with the local context. In fact, qualitative data are almost always collected through interviews and focus groups or with forms of participation usually adopted in social science disciplines. An increasing amount of studies is then focusing on the spatial aspects of the phenomenon, linking the role of the urban space with the poverty [36–39]. In fact, there are several aspects linked with the spatial approach: on the one hand, there are social connections at the neighborhood dimension that can mitigate energy poverty by creating chains of solidarity and support; on the other hand, the homogeneous district often has similar characteristics in terms of building performance and energy necessities. Thus, implementing a spatial approach into the analysis of energy poverty can give new insights both on the design/technical and social perspectives.

Finally, another interesting method for improving knowledge on energy poverty and, together, studying new strategies is strictly connected with BIM innovations. As described in Zhang et al. [40], in fact, the Building Information Technology can be successfully used also for energy poverty by adding socio-economic information to the model. This allows for investigating retrofitting solutions, among a panel of interventions, also considering poverty-related information.

Among the papers centered on methodologies, an important number of them proposes new metrics of measurement in terms of indicators or indexes. A deeper analysis of these is provided in the following paragraphs.

The second category where papers are concentrated is the theoretical one. As a matter of fact, the topic is still under analysis especially for what concerns the link between energy poverty and energy justice. Attention is in fact related to the ethical aspects of energy and to the necessity to guarantee its access to everybody [40]. However, the majority of theory-related papers are concentrated between 2015 and 2017, with less development in the last two years, where, instead, the new topic of energy citizenship arises. With energy citizenship intended, there is participation of customers in energy production not only as passive actors of the market but as prosumers and active participants [41]. However, the link between energy poverty and energy citizenship is still under-explored, and it constitutes an interesting line of research for the future of the topic.

The systematic literature review showed some red lines and some current trends of the international debate about fuel and energy poverty. The two core themes that seem to be highly relevant for the extent of this research are mainly two: metrics for increasing knowledge on the phenomenon, which is the base for then addressing the problem with efficient solutions, and profiling as a way to study solutions (products and services) specifically targeted to different categories of people.

4. A Qualitative Analysis on Three Core Aspects of Energy Poverty: Methods of Measurement, Users and Vulnerabilities, Mitigation Measures

4.1. Measuring Energy Poverty: Critiques to Standard Methods

The current multiplicity of measurement methods and the growing research interest in providing alternative metrics is a critical point. Data incompatibility, different database and data sources, different measurement units, and different scales of reference are only some of the limitations to a national/international metric for the phenomena.

A review of the most recent proposals of metrics for energy poverty highlights clearly the fragilities of the current metrics while providing possible solutions. The UK is a pioneer in research into assessing the extent of fuel poverty, but France and Germany have also developed a considerable amount of research [42,43] into the issue of assessing and identifying who is most vulnerable.

Moving forward from the initial indication of less than 10% of income as an indicator of energy poverty [31], EPOV has been providing and gathering a series of data concerning energy poverty and, more importantly, it has been collecting different metrics. This data stream has been providing more evidence-based approaches to energy poverty [12], but also a higher awareness of the multi-dimension of the problem and of its metrics. Many authors [44,45] recall the critical issues arising from geographical diversity (different countries require different mixes of policies and measures to address energy poverty) and of scale (macro level and policies and micro level to implement tailored-based measures). To this is added the lack of a common definition, one of the main causes of the insufficient policy measures adopted so far [46]. In the light of this complexity, several attempts have been made to develop a number of effective measurement strategies, covering not only diversified domestic energy uses, but also the habits of different segments of the population.

The amount of variables and indicators to be aggregated to measure energy poverty—and eventually to provide a profile of the consumer—is linked to the different interpretations and definitions of such topic [44,47]. The indicators provided have a varied nature (e.g., self-reported experiences, income, energy expenditure, building features) and rely on secondary dimension linked to energy poverty determinants (e.g., energy market, climate, cost of living). This opens up a criticality in terms of research on the topic, of political framing (related to welfare or housing or energy), and of possible actions to tackle it. A further challenge for research entails finding a key for its reading that is multi-disciplinary, related to specific context, user-cantered but at the same time that provides practical, economic, and feasible solutions to be applied short-term.

As reviewed by many articles [45,47,48], the European Commission has been suggesting since 2010 a series of consensual indicators deriving from existing data surveys. With a consensual approach to energy poverty measurement, Pye et al. [28] suggest "a pragmatic approach" tailored according to each member state according to their best available data. Nevertheless, this attempt seems insufficient to deal with the lack of correlation among the definition of energy poverty, the clear identification of the users and the indicators to monitor it. To address this multiplicity, Rademaekers et al. [49] recommend using an array of indicators in combination, such as: "high share of energy expenditure in income: part of population with share of energy expenditure in income more than twice the national median (EPOV, 2010 HBS); hidden energy poverty: part of population whose absolute energy expenditure is below half the national median (EPOV, 2010); inability to keep home adequately warm: based on self-reported thermal discomfort (Eurostat, 2016 SILC); arrears on utility bills: based on households' self-reported inability to pay utility bills on time in the last 12 months (Eurostat, 2016 SILC)" [8]. This multi-dimension of indicators is enabled by the "Statistics on Income and Living Conditions" (Eurostat-SILC) that provides a set of proxy indicators used to compare energy poverty within the EU. A multi-indicators approach is also suggested by Herrero [48] who calls for methods that might capture as many variables as possible, not only in terms of domestic energy services measurement but also intersecting social, behavioral, demographic, etc. The author warns that a single-indicator might reduce the vastity and multi-disciplinarily of the issue and therefore of the possibility to tackle it. He urges the use of several indicators able "to capture the diversity of experiences and intensities" [48] of the issue.

These methodologies might be useful to frame energy poverty in a specific moment and place. Nevertheless, the phenomenon is subject to fluctuations and variations related to primary and secondary factors. Furthermore, data are referred either to the production sectors or to the type of energy carrier or source, such as, for example, the production or distribution of electricity, the need for electricity or heat. Moreover, the data may come from different sources: data collected and managed by institutional entities, authorities or other public or private research bodies, or open data. The different databases collect and return the information following their own criteria, in different format and aggregation. This kind of database on energy needs or consumption for the various sectors are not directly comparable with other databases, e.g., for the construction and real estate sector or the demographic sector. In other words, there is no direct correlation between energy needs and consumption and their identification and location at urban level.

For this reason, Rademaekers et al. [49] provide an additional set of indicators (demographic factors, energy prices, income, kind of household, heating system, supply choice, building efficiency and building stock, policy intervention) to be applied to cross-identify it.

This short review of the measurements methodologies shows a large effort in the quantification of the phenomena, in spite of a discursive and narrative approach, based more on quality. Despite showing a common denominator in the observation of the individual user, the approaches lack a narrative, explorative, and descriptive dimension and risk to exclude some categories and variables from the analysis. The user role remains underexplored; nevertheless, its key importance is widely mentioned, using different labels and evoking concepts. This reading key is addressed in the following chapter.

4.2. The User Dimension of Energy Poverty

To tackle energy poverty implies a change of paradigm in the everyday lifestyles and a different configuration of the urban socio-technical systems, linked to energy as a social necessity [50]. These systems (infrastructural, electric, logistic, waste, etc.), in fact, require a complete review in the way they are organised, distributed, and managed. Both their technical features and their social roles and responsibilities [50] need to change in order to provide the basis for the support of the transition towards a more efficient and sustainable urban system. In the research on energy poverty, the key role of users has been largely recalled by literature. Energy saving linked to user behavior is deemed as significant as that deriving from technology [51]. Two main orientations towards the users are highlighted: target users as "vulnerable customers" and pro-active users indicated as "energy communities" and "energy citizenship".

The concept of vulnerable customers can include income levels, the available share of energy expenditure, energy efficiency of households, dependence on electrical equipment for health reasons, age, among other criteria. Pye et al. [28] categorized the different interpretation of the concept according to each Member State, highlighting the recurrence of a user as a receipt of social welfare, while in France, Sweden, and Italy, it reflects the relationship between low income and high expenditure. Recent studies include vulnerable users, low-income families with children, elderly people in nursing homes, rented persons, and those with an unstable employment situation [52,53]. Directive 2019/944 of 5 June 2019 concerning common rules for the internal market in electricity and amending Directive 2012/27/EU, in Article 28 identifies users as "vulnerable customers", stating that "member States shall take appropriate measures to protect customers and shall in particular ensure that vulnerable customers are adequately protected". Around this concept, a working group of the European Commission was created, following the 2015 Citizen Energy Forum: the Vulnerable Consumer Working Group (VCWG), in charge of exploring the potential for common approaches across the EU to vulnerability and energy poverty users' definitions and policies. This concept, however, seems to narrow the field of application to mere economic customers' communities, identifying them with economic/financial indicators and moving away from the concept of energy as a necessity and a human right [54], but rather towards energy as a commodity. An attempt to overcome this risk is to face the issue extending the idea of energy to the concept of citizenship meant as collective ownership and exercise of rights.

Designing a different role for the users, it is possible to imagine a proactive community, including vulnerable groups, that progressively adopts effective solutions, strategies, and management policies to face collectively the energy poverty issues.

The concept of energy citizenship [34,55–57] entails an "approach [that] argues for the social necessity of public engagement and participation in processes of policy-making and planning, driven by principles of local empowerment" [55]. This definition urges the citizens to be involved not only in a personal change of their behavior—by reducing their energy consumption expenditures (EU Directive 2018/2012)—but also in being agents and providers of solutions to address (part of) the wider energy poverty problem. In this view, users are called to "shape new routines and enact system change" [58]; therefore, they seem to be autonomously drawn to behave in order to improve their condition while at the same time set an example for a deep change toward a more efficient energy consumption and use. Ultimately, they should be politically influential [56], hence acting as social and political actors [34,35].

Many authors argued that this interpretation of "user as energy citizen" tends to "ignore crucial questions of unequal agency and access to resources" [50] identifying the citizens only as those who can purchase and spend in a market regime, causing exclusion issues and failing to include the same categories affected by energy poverty issues, and the vulnerable customers themselves. Bues and Gailing [59] stress the significative role of economic power dynamics in citizens' engagement, which limit the access to technology and economic/financial resources, as well as to knowledge, information and data. Lennon et al. [50] propose to "re-conceptualize energy citizenship, moving away from individualist and economistic perspectives and locating it within collective contexts of engagement". This approach seems to recall what has been commonly defined as energy communities, another evocative label used to frame not only the attention to the user, but its pro-active role in the management and improvement of sustainable behaviors around energy.

The energy communities are introduced in the Regulation (EU) 2018/1999 (11/12/2018), in particular in the art. 20 "Integrated Communications on renewable energy" as "the subjects to be promoted and facilitated for the self-consumption of renewable energy" (letter b paragraph 7). This definition followed the growing decentralization and distribution energy systems [33,60] as a precondition likely to produce a new category of actors engaged in the energy system with new active roles.

In the preamble of Regulation 2018/1999, it is pointed out that "[...] member States should assess the number of households in energy poverty, taking into account the domestic energy services needed to ensure a basic standard of living in their national context, existing social policy and other relevant policies, as well as the Commission's indicative guidelines on related indicators, including geographical dispersion, which are based on a common approach to energy poverty".

It is clear that the attention to the user (as vulnerable customer) and its pro-active involvement as agent—energy community, energy citizenship—is key for tackling energy poverty issues in Europe. Nevertheless, some authors [34] highlight the lack of practical indication of measures or stress the limitation in these general labels. In fact, these approaches tend to overlook the qualitative identification criteria of such agents, their behavioral patterns, their possibility to access resources (in terms of knowledge, information, as well as financial ones) together with the connection with the characteristics of the buildings in which they live and the correlation with their demographic and financial situation. This opens up a reflection on accessibility, to information and knowledge. European cities and metropolis particularly benefit from the presence of knowledge structures, the density and easiness of connection and access to information and economic sources. The territorial reality in which the cities fit, however, presents phenomena of increasing social and political polarization, depopulation of internal territories and the removal of disadvantaged social classes in favor of a few categories that hold most of the knowledge capital. In this scenario, it appears that the majority of the poorer energy communities may be left out from the decision-making epicenters, as well as from the possibility of political participation and representation.

It is widely known that energy poverty emerges as determined by three factors: income, energy cost, and building characteristics. Following the above-mentioned strategies that focus on the role of the user as consumer and producers of solutions, the challenge is to adopt a new design approach, able to combine spatial issues with new services, new enabling technologies, and new supportive management tools.

4.3. Mitigation Strategies from Literature

Participation is one of the ways in which energy poverty can be mitigated. Some studies indeed focus on the role of community participation in improving the knowledge on energy and on energy services. As an example, Martiskainen et al. [61,62] describe the organization of Energy Cafés, a format for letting communities engaging with energy locally. In particular, the community organize pop-up temporary events where information and advice on energy are provided to other persons. As assessed in the conclusion of the paper, "Energy Cafés open up for various forms of advocacy, highlighting a broken link between the expectations of the energy markets and energy practices in the home" [62].

Participation is also intended as the involvement of other actors besides end-users (which usually are residents). Several studies [57,63] focus in fact on the benefits of involving in a coordinated analysis and problem-solving other categories of persons, such as owners and tenants, social workers, and healthcare practitioners.

Policy-related papers (16% of the total) are mainly focused on the analysis of current policies in different European countries. Especially addressed to analyze the UK situation, some others compare different countries among them.

Profiling-related papers (12%) are of particular interest as they recognize how energy poverty can be different in relation with different people living in a household. In fact, there are several categories of people potentially more exposed to energy poverty, but also energy poverty can affect them differently (also with different health effects). In relation with the literature review, actually, the categories of persons more studied in relation with energy poverty are students, elders and young. However, up to now, it seems that research is concentrated in identifying the effects of fuel poverty on them instead of defining solutions specifically designed for these categories. This last point doesn't seem to be present in most of the papers, but it constitutes an interesting topic to be further developed.

From a technical point of view, 10% of analyzed papers focus on technical and technological solutions to fuel and energy poverty. Among them, most are based on finding solutions to produce more energy from renewable sources [64–68]. In particular, Donaldson et al. [68] propose the use of urban brownfields for producing energy targeted to poor households. Some research is then focused on the implementation of storage systems and lithium batteries together with energy production [64,65]. Very few papers focus on retrofitting solutions as a way to reduce and mitigate energy poverty.

Often, researchers argue that deep retrofitting solutions are costly and they usually need a temporary relocation of tenants. However, as in Aranda et al. [69], some solutions are proposed, such as working on the building envelopes, on lighting systems and on energy generation. Nevertheless, less disruptive and low-cost retrofitting solutions (also not linked to solving energy poverty) have been studied by architectural and engineering researchers for years, but they do not seem to constitute the unique response to fuel poverty because, as many authors argue, fuel and energy poverty also involves social, political, and economic concerns [70–74].

5. Discussion and Conclusions

The systematic literature review showed how currently it is possible to recognize some red lines in which the European research on fuel and energy poverty is mostly aligned.

At first, most research is still investigating the knowledge about the topic, in terms of definitions, semantic analysis, and theories, but also in terms of collecting and analyzing data. For the matter of data collection, it seems that the current research is mostly aligned to merge quantitative data coming from official sources (local, national, or EU) with qualitative data coming from different forms of participatory approaches (from interviews, focus groups, and observation).

A high amount of effort is put into defining the most suitable indexes and metrics to describe the phenomenon. In terms of technical solutions, most of the literature is aligned in working mainly on RES, storage, and less in retrofitting solutions. However, the role of technical solutions doesn't seem to be enough to mitigate the problem, as poverty has deep social, political, and economic roots that need to be understood.

For that reason, several studies are now proposing the role of urban actors and solidarity chains at the neighborhood level as a way to complement, with soft solutions, the harder component of retrofitting.

Most studies, then, are concentrated in the United Kingdom, as one of the first countries highly engaged in addressing the issue, even if some other countries are showing an increasing amount of attention, especially Spain and, in general, the Southern European countries.

Despite the presence of a high interest in the topic, some barriers and limitations can still be found, as well as potentially interesting new lines of research.

The limitations and barriers have been distinguished in three categories: economic, social, and technic/technological.

Concerning the economic issues, the high cost of the energy mitigation measures of the single or block of buildings is still one of the major challenges. As, for example, indicated in the Report on the state of energy poverty in Italy (2018), climate change affects household energy demand, exerting upward pressure on energy prices, whose incidence on the total rose from 4.7 per cent in 2007 to 5.1 per cent in 2017. The percentage incidence of energy expenditure is higher for less well-off households (those with a lower than average expenditure). The condition of these households has worsened over the last decade: in 2007, 20% of the least affluent households spent about 6% of total spending on lighting, heating, cooking, and cooling domestic environments, and, ten years later, this share increased by about half a percentage point (while it was stable or even decreasing for other households). In fact, electricity prices have long been burdened by systemic charges linked to the support of renewable energy and the mechanism for pricing emissions within the EU; in the future, the same will happen for other energy sources that will be subject either to some form of carbon tax or to other forms of restrictions on their use (such as the use of coal in electricity generation).

According to many predictions [22,45], European cost of electricity will rise further in the coming decades, due to the energy transition, the prosecution of the economic crisis and a serious de-carbonization struggle that will lead to a further increase in energy prices.

A second economic issue regards the difficulties in systemically eradicating the issue with subsidies and economic support measure. As pointed out by Bouzarovski [54], in fact, subsidies and measures that avoid disruption as a result of repeated arrear payments temporarily alleviate the energy poverty of households, but do not solve the problem at all.

The social aspects concern understanding and awareness, understood as self-awareness or understanding of the phenomenon of the population regarding energy and energy efficiency. Certain conditions such as energy poverty are not understood as emergency conditions by the population. While people who are in the condition of Absolute Poverty are identified and aware and have access to social services in the territory (Municipality, Caritas, etc.), the condition of Energy Poverty is not recognized.

Many financial schemes and social benefits have been developed to support the installation of measures that reduce energy consumption [36], but they are not well exploited, as low-income households cannot save the necessary funds to cover their initial expenses and generally have difficulty in obtaining a loan.

The condition of Energy Poverty involves a series of actions by subjects that can be "measured" in social/sociological terms, such as:

- difficulty in paying bills;
- management of how to use heating systems (turning the system off and staying in the cold to spend less);
- tolerating conditions of discomfort, hot/cold, for long periods, particularly in summer (heat waves, etc.).
- health conditions, pathologies, etc.

Indicators of social aspects should be oriented to the identification of the above-mentioned conditions. Nevertheless, vulnerable groups of citizens are not easily categorized because their individual needs, knowledge, culture, or living conditions cannot always be simply labelled. Therefore, knowledge transfer on energy use and energy efficiency advice should be targeted to people's specific social situations. The variety of stress factors and living conditions makes knowledge transfer difficult.

The technical/technological limitations do not specifically concern knowledge or development limitations, but rather expensive applications of them. Household efficiency measures are particularly relevant for reducing energy poverty, but much can also be done in the direction of energy savings in other areas. Low cost solutions [8] might be a temporary answer to tackle emergencies (e.g., monitoring systems, room control temperature, devices for the production of DHW—Domestic Hot Water, selection of energy-efficient installations and equipment, electrical storage batteries). Additionally, as the major signs of energy poverty within people are difficultly detectable (e.g., low indoor temperature), it seems that the current progressive transition to smart cities, and the use of digital connected devices in households—such as smart thermostats—can be of great help for gaining a deeper insight also on the energy poverty topic.

Among the solutions proposed to support the increase of energy efficiency in housing are related to:

- norms and regulations: mandatory construction standards for new constructions (e.g., insulation materials, windows); mechanisms that provide for the installation of efficient heating systems/electrical installations by service providers and whose costs are repaid by consumers over time through the supply contract;
- subsidiary measures: energy efficiency improvements in the home, sometimes with higher amounts for lower incomes; reduce the costs of the investment (i.e., tax deduction for part of the expenses for efficiency);
- educational: establishment of "energy tutors" who advise, on a case-by-case basis, what choices to make to improve the energy efficiency of the home;
- specific actions for homeowners: targeted actions to qualify the investment made (issue of energy efficiency certificates following the intervention, profits for the rental of the property).
- welfare and policy: including better social tariff, new services, energy audit, investment assistance.

Another interesting aspect, emerging from the literature, is the necessity to link data with the spatial dimension by using the technologies of BIM and GIS. The use of such software can in fact be helpful in understanding the urban dynamics on energy poverty and, thus, in finding specific solutions to each context. However, it is necessary to understand the availability of qualitative data from the public and energy providers and to boost their transition to open source aggregated datasets.

In conclusion, it seems that the problem needs an integrated and holistic approach, able to combine the technical issues about building performances, enabling technologies, with a deeper understanding of social connections, economic conditions (also in terms of investigating why people are in economic difficulties). A complete understanding of people behavior and awareness can help in the characterization of profiles to face the several challenges connected to energy poverty.

Behaviors and local cultural factors can drive basic energy use practices [75]: the factors and their relationships that influence consumer behavior and practices are dynamic, highly dependent on human elements, change over time, and influence consumer behavior. For these regions, the process of consumption practices becomes unpredictable and detached from any recognizable pattern [76]. Investigating the size and value of end-user profiles and behaviors before the design phase [77] helps to identify the best combination of technical and social measures, through the adoption of user-friendly energy efficiency systems that could reduce the energy poverty, facilitate the use by tenants and increase their environmental and energy awareness.

The design challenge is to contribute to the necessary framework made of technical and social solutions, policies for energy justice, and participatory processes to understand the energy equity dimension in terms of accessibility and affordability [39].

Limitations and Further Research

The paper has some limitations due to its choices and methodology. At first, the decision to exclude non-English written papers resulted, obviously, in a limitation in the number of analyzed findings. Especially for what concern locally specific case studies, some interesting approaches were probably not included into the analysis for that reason. However, we believe that the general analysis and the major findings of the paper have not been penalized by this choice.

The paper identified some interesting trends that could be deepened in future research and, in particular, the following:

- the connection between energy poverty and the spatial approach that use BIM and GIS software allowing a coordinated approach among technicians and the geo-localization of vulnerabilities seems to be very interesting in order to deeply understand the phenomenon and its spatial specificities;
- the necessity to study the presence of potentially different profiles of vulnerable people and thus to study services more aligned to their specificities.

Author Contributions: Conceptualization, A.B. and V.G.; methodology, A.B. and V.G.; formal analysis, S.O.M.B. and M.M.; investigation, S.O.M.B. and M.M.; resources, S.O.M.B. and M.M.; data curation, S.O.M.B. and V.G.; writing—original draft preparation, S.O.M.B. and M.M.; writing—review and editing, A.B. and V.G.; visualization, S.O.M.B.; supervision, A.B. and V.G.; project administration, A.B and S.O.M.B.; funding acquisition, A.B. All authors have read and agreed to the published version of the manuscript.

Funding: This research received no external funding.

Conflicts of Interest: The authors declare no conflict of interest.

References

1. Bouzarovski, S.; Petrova, S.; Sarlamanov, R. Energy poverty policies in the EU: A critical perspective. *Energy Policy* **2012**, *49*, 76–82. [CrossRef]
2. Bouzarovski, S. Energy poverty in the European Union: Landscapes of vulnerability. *Wiley Interdiscip. Rev. Energy Environ.* **2014**, *3*, 276–289. [CrossRef]

3. Bouzarovski, S.; Petrova, S. A global perspective on domestic energy deprivation: Overcoming the energy poverty–fuel poverty binary. *Energy Res. Soc. Sci.* **2015**, *10*, 31–40. [CrossRef]

4. Simcock, N.; Thomson, H.; Petrova, S.; Buzar, S. *Energy Poverty and Vulnerability: A Global Perspective*; Routledge: London, UK, 2018.

5. Birol, F. Energy economics: A place for energy poverty in the agenda? *Energy J.* **2007**, 28. [CrossRef]

6. Nussbaumer, P.; Bazilian, M.; Modi, V. Measuring energy poverty: Focusing on what matters. *Renew. Sustain. Energy Rev.* **2012**, *16*, 231–243. [CrossRef]

7. What is Energy Poverty? Available online: https://www.energypoverty.eu/about/what-energy-poverty (accessed on 10 January 2020).

8. EnR Position Paper on Energy Poverty in the European Union—January 2019. Available online: http://enr-network.org/wp-content/uploads/ENERGYPOVERTY-EnRPositionPaper-Energypoverty-Jan-2019.pdf (accessed on 27 December 2019).

9. Thomson, H.; Snell, C.J.; Liddell, C. Fuel poverty in the European Union: A concept in need of definition? *People Place Policy Online* **2016**, 5–24. [CrossRef]

10. EC–European Commission. *Accelerating Clean Energy Innovation: Clean Energy for All Europeans*; Publications Office: Brussels, Belgium, 2016.

11. Bailis, R. *Energy and Poverty: The Perspective of the Poor Countries, Handbook of Sustainable Energy*; Edward Elgar: Cheltenham, UK, 2012.

12. Thomson, H.; Snell, C.; Bouzarovski, S. Health, well-being and energy poverty in Europe: A comparative study of 32 European countries. *Int. J. Environ. Res. Public Health* **2017**, *14*, 584. [CrossRef]

13. Pye, S.; Dobbins, A.; Baffert, C.; Brajković, J.; Deane, P.; De Miglio, R. Addressing Energy Poverty and Vulnerable Consumers in the Energy Sector Across the EU. *L'Europe Form.* **2015**, *378*, 64–89. [CrossRef]

14. Covenant of Mayors. Available online: https://www.eumayors.eu/support/energy-poverty.html (accessed on 10 January 2020).

15. Heffner, G.; Campbell, N. Evaluating the co-benefits of low-income energy-efficiency programmes. In *Workshop Report*; OECD/IEA: Paris, France, 2011.

16. *The Sustainable Development Goals Report*; United Nations: New York, NY, USA, 2018; Available online: https://unstats.un.org/sdgs/files/report/2018/TheSustainableDevelopmentGoalsReport2018-EN.pdf (accessed on 10 January 2020).

17. *Environmental Performance Index*; EPI Report; Yale University: New Haven, CT, USA, 2018; Available online: https://epi.envirocenter.yale.edu/2018-epi-report/introduction? (accessed on 17 December 2019).

18. Faiella, I.; Lavecchia, L. La povertà energetica in Italia. *Politica Econ.* **2015**, *31*, 27–76.

19. Papada, L.; Kaliampakos, D. A Stochastic Model for Energy Poverty Analysis. *Energy Policy* **2018**, *116*, 153–164. [CrossRef]

20. European Commission. *Clean Energy for all Europeans: Commission welcomes European Parliament's Adoption of New Electricity Market Design Proposals*; Press Release: Brussels, Belgium, 2019. Available online: http://europa.eu/rapid/press-release_IP-19-1836_en.htm (accessed on 10 January 2020).

21. Dubois, U.; Meier, H. Energy affordability and energy inequality in Europe: Implications for policymaking. *Energy Res. Soc. Sci.* **2016**, *18*, 21–35. [CrossRef]

22. Santillán Cabeza, S.E. Opinion of the European Economic and Social Committee on Energy poverty in the Context of Liberalisation and the Economic Crisis. 2010. Available online: https://www.energypoverty.eu/publication/opinion-european-economic-and-social-committee-energy-poverty-context-liberalisation-and (accessed on 10 January 2020).

23. Dubois, U. From targeting to implementation: The role of identification of fuel poor households. *Energy Policy* **2012**, *49*, 107–115. [CrossRef]

24. Palmer, G.; MacInnes, T.; Kenway, P. Cold and Poor: An Analysis of the Link between Fuel Poverty and Income, New Policy Institute & EAGA Charitable Trust. 2008. Available online: http://www.poverty.org.uk/reports/fuel%20poverty.pdf (accessed on 10 January 2020).

25. Bouzarovski, S.; Tirado, H.S. The energy divide: Integrating energy transitions, regional inequalities and poverty trends in the European Union. *Eur. Urban Reg. Studies* **2017**, *24*, 69–86. [CrossRef] [PubMed]

26. Bridge, G.; Bouzarovski, S.; Bradshaw, M.; Eyre, N. Geographies of energy transition: Space, place and the low-carbon economy. *Energy Policy* **2013**, *53*, 331–340. [CrossRef]

27. Cecelski, E. Enabling equitable access to rural electrification: Current thinking on energy, poverty, and gender. In *Briefing Paper, Asia Alternative Energy Policy and Project Development Support*; World Bank: Washington, DC, USA, 2002.

28. Pye, S.; Dobbins, A.; Baffert, C.; Brajković, J.; Deane, P.; De Miglio, R. Energy poverty across the EU: Analysis of policies and measures. In *Europe's Energy Transition-Insights for Policy Making*; Academic Press: Cambridge, MA, USA, 2017.

29. Primc, K.; Slabe-Erker, R.; Majcen, B. Constructing energy poverty profiles for an effective energy policy. *Energy Policy* **2019**, *128*, 727–734. [CrossRef]

30. Siddaway, A.P.; Wood, A.M.; Hedges, L.V. How to do a systematic review: A best practice guide for conducting and reporting narrative reviews, meta-analyses, and meta-syntheses. *Annu. Rev. Psychol.* **2019**, *70*, 747–770. [CrossRef]

31. Boardman, B. *Fuel Poverty: From Cold Homes to Affordable Warmth*; Belhaven Press: London, UK, 1991.

32. Bouzarovski, S. *Energy Poverty. (Dis)Assembling Europe's Infrastructural Divide*; Palgrave MacMillan: Cham, Switzerland, 2018.

33. Horta, A.; Gouveia, J.P.; Schmidt, L.; Sousa, J.C.; Palma, P.; Simões, S. Energy Poverty in Portugal: Combining Vulnerability Mapping with Household Interviews. *Energy Build.* **2019**, *203*, 109423. [CrossRef]

34. Petrova, S. Illuminating Austerity: Lighting Poverty as an Agent and Signifier of the Greek Crisis. *Eur. Urban Reg. Stud.* **2017**, *25*, 360–372. [CrossRef]

35. Chavez, A.C.D.; Gilbertson, J.; Tod, A.M.; Nelson, P.; Powell-Hoyland, V.; Homer, C.; Lusambili, A.; Thomas, B. Using Environmental Monitoring to Complement in-Depth Qualitative Interviews in Cold Homes Research. *Indoor Built Environ.* **2017**, *26*, 937–950. [CrossRef]

36. Gouveia, J.P.; Palma, P.; Simoes, S.G. Energy Poverty Vulnerability Index: A Multidimensional Tool to Identify Hotspots for Local Action. *Energy Rep.* **2019**, *5*, 187–201. [CrossRef]

37. Martin-Consuegra, F.; Hernandez-Aja, A.; Oteiza, I.; Alonso, C. Distribution of energy poverty in the city of Madrid (Spain). *Eure-Rev. Latinoam. De Estud. Urbano Reg.* **2019**, *45*, 133–152. [CrossRef]

38. Robinson, C.; Lindley, S.; Bouzarovski, S. The Spatially Varying Components of Vulnerability to Energy Poverty. *Ann. Am. Assoc. Geogr.* **2019**, *109*, 1188–1207. [CrossRef]

39. Bouzarovski, S.; Simcock, N. Spatializing Energy Justice. *Energy Policy* **2017**, *107*, 640–648. [CrossRef]

40. Zhang, Y.; Tzortzopoulos, P.; Kagioglou, M. Healing Built-Environment Effects on Health Outcomes: Environment–Occupant–Health Framework. *Build. Res. Inf.* **2018**, *47*, 747–766. [CrossRef]

41. Sovacool, B.K.; Heffron, R.J.; Mccauley, D.; Goldthau, A. Energy Decisions Reframed as Justice and Ethical Concerns. *Nat. Energy* **2016**, *1*. [CrossRef]

42. ONPE. Premier rapport de l'ONPE—Observatoire National de la PrécaritéEnergétique, Définitions, Indicateurs, Premiers Résultats et Recommandations. September 2014. Available online: https://onpe.org/sites/default/files/pdf/documents/rapports_onpe/onpe1errapportsynthese.pdf (accessed on 10 January 2020).

43. Schreiner, N. Auf der Suche nach Energiearmut: Eine Potentialanalyse des Low-Income-High-Costs Indikators für Deutschland. SOEP Papers on Multidisciplinary Panel Data Research at DIW Berlin. Available online: https://www.diw.de/de/diw_01.c.523076.de/publikationen/soeppapers/2015_0811/auf_der_suche_nach_energiearmut__eine_potentialanalyse_des_low-income-high-cost_indikators_fuer_deutschland.html (accessed on 10 January 2020).

44. Thomson, H.; Bouzarovski, S.; Snell, C. Rethinking the measurement of energy poverty in Europe: A critical analysis of indicators and data. *Indoor Built Environ.* **2017**, *26*, 879–901. [CrossRef]

45. Faiella, I.; Lavecchia, L. Energy poverty indicators. In *Urban Fuel Poverty*; Academic Press: Cambridge, MA, USA, 2019.

46. Santangelo, A.; Yan, D.; Feng, X.; Tondelli, S. Renovation strategies for the Italian public housing stock: Applying building energy simulation and occupant behavior modelling to support decision-making process. *Energy Build.* **2018**, *167C*, 269–280. [CrossRef]

47. Bollino, C.A.; Botti, F. Energy poverty in Europe: A multidimensional approach. *PSL Q. Rev.* **2017**, *70*, 473–507.

48. Herrero, S.T. Energy poverty indicators: A critical review of methods. *Indoor Built Environ.* **2017**, *26*, 1018–1031. [CrossRef]

49. Rademaekers, K.; Yearwood, J.; Ferreira, A.; Pye, S.; Hamilton, I.; Agnolucci, P.; Anisimova, N. *Selecting Indicators to Measure Energy Poverty*; Trinomics: Rotterdam, The Netherlands, 2016.

50. Lennon, B.; Dunphy, N.; Gaffney, C.; Revez, A.; Mullally, G.; O'Connor, P. Citizen or consumer? Reconsidering energy citizenship. *J. Environ. Policy Plan.* **2019**, 1–14. [CrossRef]

51. Lopes, M.A.R.; Antunes, C.H.; Martins, N. Energy behaviors as promoters of energy efficiency: A 21st century review. *Renew. Sustain. Energy Rev.* **2012**, *16*, 4095–4104. [CrossRef]

52. Romero, J.C.; Linares, P.; López, X. The policy implications of energy poverty indicators. *Energy Policy* **2018**, *115*, 98–108. [CrossRef]

53. Berry, A. The distributional effects of a carbon tax and its impact on fuel poverty: A microsimulation study in the French context. *Energy Policy* **2019**, *124*, 81–94. [CrossRef]

54. Bouzarovski, S. Energy Poverty Policies at the EU Level. In *Energy Poverty*; Palgrave Macmillan: Cham, Switzerland, 2018.

55. Devine-Wright, P. Energy citizenship: Psychological aspects of evolution in sustainable energy technologies. In *Governing Technology for Sustainability*, 1st ed.; Murphy, J., Ed.; Routledge: London, UK, 2007; pp. 63–88.

56. Devine-Wright, P. Reconsidering Public Attitudes and Public Acceptance of Renewable Energy Technologies: A Critical Review. Available online: http://geography.exeter.ac.uk/beyond_nimbyism/deliverables/Reconsidering_public_acceptance.pdf (accessed on 10 January 2020).

57. McConalogue, D.; Kierans, C.; Moran, A. The hidden practices and experiences of healthcare practitioners dealing with fuel poverty. *J. Public Health* **2016**, *38*, 206–211. [CrossRef]

58. Schot, J.; Kanger, L.; Verbong, G. The roles of users in shaping transitions to new energy systems. *Nat. Energy* **2016**, *1*, 16054. [CrossRef]

59. Bues, A.; Gailing, L. Energy transitions and power: Between governmentality and depoliticization. In *Conceptualizing Germany's Energy Transition*; Gailing, L., Moss, T., Eds.; Palgrave Pivot: London, UK, 2016.

60. Ryghaug, M.; Skjølsvold, T.M.; Heidenreich, S. Creating Energy Citizenship through Material Participation. *Soc. Stud. Sci.* **2018**, *48*, 283–303. [CrossRef]

61. Martiskainen, M.; Speciale, G. 'The Fuel Bill Drop Shop': An Investigation into Community Action on Fuel Poverty; Final Report to Chesshire Lehmann Fund; South East London Community Energy (SELCE): London, UK, 2016; Available online: http://tinyurl.com/huxverc (accessed on 10 January 2020).

62. Martiskainen, M.; Heiskanen, E.; Speciale, G. Community energy initiatives to alleviate fuel poverty: The material politics of Energy Cafés. *Local Environ.* **2018**, *23*, 20–35. [CrossRef]

63. Scarpellini, S.; Hernández, M.A.S.; Llera-Sastresa, E.; Aranda, J.A.; Rodríguez, M.E.L. The mediating role of social workers in the implementation of regional policies targeting energy poverty. *Energy Policy* **2017**, *106*, 367–375. [CrossRef]

64. Paniyil, P.; Singh, R.; Asif, A.; Powar, V.; Bedi, G.; Kimsey, J. Transformative and disruptive role of local direct current power networks in power and transportation sectors. *Facta Univ. Ser. Electron. Energetics* **2019**, *32*, 387–402. [CrossRef]

65. Yu, Y.; Narayan, N.; Vega-Garita, V.; Popovic-Gerber, J.; Qin, Z.; Wagemaker, M.; Bauer, P.; Zeman, M. Constructing Accurate Equivalent Electrical Circuit Models of Lithium Iron Phosphate and Lead–Acid Battery Cells for Solar Home System Applications. *Energy* **2018**, *11*, 2305. [CrossRef]

66. Papadopoulou, S.D.; Kalaitzoglou, N.; Psarra, M.; Lefkeli, S.; Karasmanaki, E.; Tsantopoulos, G. Addressing Energy Poverty through Transitioning to a Carbon-Free Environment. *Sustainability* **2019**, *11*, 2634. [CrossRef]

67. Strielkowski, W.; Volkova, E.; Pushkareva, L.; Streimikiene, D. Innovative Policies for Energy Efficiency and the Use of Renewables in Households. *Energies* **2019**, *12*, 1392. [CrossRef]

68. Donaldson, R.; Lord, R. Can brownfield land be reused for ground source heating to alleviate fuel poverty? *Renew. Energy* **2018**, *116*, 344–355. [CrossRef]

69. Aranda, J.; Zabalza, I.; Conserva, A.; Millán, G. Analysis of energy efficiency measures and retrofitting solutions for social housing buildings in Spain as a way to mitigate energy poverty. *Sustainability* **2017**, *9*, 1869. [CrossRef]

70. DellaValle, N. People's decisions matter: Understanding and addressing energy poverty with behavioral economics. *Energy Build.* **2019**, *204*, 109515. [CrossRef]

71. Middlemiss, L. A critical analysis of the new politics of fuel poverty in England. *Crit. Soc. Policy* **2017**, *37*, 425–443. [CrossRef]

72. Longhurst, N.; Hargreaves, T. Emotions and fuel poverty: The lived experience of social housing tenants in the United Kingdom. *Energy Res. Soc. Sci.* **2019**, *56*, 101207. [CrossRef]

73. Kearns, A.; Whitley, E.; Curl, A. Occupant behavior as a fourth driver of fuel poverty (aka warmth & energy deprivation). *Energy Policy* **2019**, *129*, 1143–1155. [CrossRef]
74. Webb, J.; Hawkey, D.; McCrone, D.; Tingey, M. House, home and transforming energy in a cold climate. *Fam. Relatsh. Soc.* **2016**, *5*, 411–429. [CrossRef]
75. IPCC. Climate Change. Mitigation of climate change. In *Contribution of Working Group III to the Fifth Assessment Report of the Intergovernmental Panel on Climate Change*; Cambridge University Press: New York, NY, USA, 2014.
76. European Environment Agency. *EMEP/EEA Air Pollutant Emission Inventory Guidebook*; No 12/2013; Publication Office of the European Union: Luxembourg, 2013.
77. Gianfrate, V.; Piccardo, C.; Longo, D.; Giachetta, A. Rethinking social housing: Behavioral patterns and technological innovations. *Sustain. Cities Soc.* **2017**, *33*, 102–112. [CrossRef]

Review

Outdoor Wellbeing and Quality of Life: A Scientific Literature Review on Thermal Comfort

Ernesto Antonini *, Vincenzo Vodola, Jacopo Gaspari and Michaela De Giglio

Department di Architecture—DA, University of Bologna, Viale Risorgimento 2, 40136 Bologna, Italy;
vincenzo.vodola2@unibo.it (V.V.); jacopo.gaspari@unibo.it (J.G.); michaela.degiglio@unibo.it (M.D.G.)
* Correspondence: ernesto.antonini@unibo.it

Received: 17 March 2020; Accepted: 9 April 2020; Published: 21 April 2020

Abstract: While indoor comfort represents a widely investigated research topic with relation to sustainable development and energy-demand reduction in the built environment, outdoor comfort remains an open field of study, especially with reference to the impacts of climate change and the quality of life for inhabitants, particularly in urban contexts. Despite the relevant efforts spent in the last few decades to advance the understanding of phenomena and the knowledge in this specific field, which obtained much evidence for the topic's relevance, a comprehensive picture of the studies, as well as a classification of the interconnected subjects and outcomes, is still lacking. This paper reports the outcomes of a literature review aimed at screening the available resources dealing with outdoor thermal comfort, in order to provide a state-of-the-art review that identifies the main topics focused by the researchers, as well as the barriers in defining suitable indexes for assessing thermal comfort in outdoor environments. Although several accurate models and software are available to quantify outdoor human comfort, the evocated state of mind of the final user still remains at the core of this uncertain process.

Keywords: outdoor thermal comfort; human thermal perception; thermal comfort assessment; quality of life

1. Introduction

1.1. Review Contest and Boundaries

While indoor conditions have been the main concern for research on user comfort since the second half of 20th century, assessing outdoor comfort has emerged as a challenging field during the last few decades. Three main phenomena have pushed towards this change:

- The growth of cities, driven by the increasing movement of people to urban areas, where half of the world's population is already living, and a further expansion is expected in the near future [1].
- The consequent exposure of a huge number of people to the effects of extreme weather conditions due to both climate change and local phenomena, boosted by the high density of settlements, such as Urban Heat Islands (UHI) (peaks of temperature higher than that of the rural surroundings) [2]. The evidence on the average temperature increase and the related potential impacts are widely explored in authoritative reports from the Intergovernmental Panel on Climate Change (IPCC) [3] and the National Oceanic and Atmospheric Administration (NOAA) [4], particularly dealing with more relevant effects on urban areas [5].
- The change in lifestyles and particularly the increasing amount of time spent by inhabitants inside buildings pushed the need for high-quality outdoor spaces that provide healthy leisure facilities, and significantly contribute to the urban environment's livability and vitality. Thus, encouraging more people to use outdoor spaces would bring greater benefits into the physical, environmental, economic and social spheres of the cities [6–9].

Scientists worldwide have thus focused their attention on this topic, making available a wide range of tools and methods to assess human thermal outdoor comfort in different climatic contexts. Over 100 biometeorological and thermal stress indexes [10] have been developed, adopting different approaches and rationales, aiming at linking the microclimatic conditions to the perceived sensations.

Since the available knowledge on human thermal perception and related evaluation protocols were mainly the ones previously developed for interior spaces and other confined spaces, the assessment of outdoor conditions initially refers to these patterns [11].

In fact, human thermal comfort and its assessment were studied since the beginning of the 20th Century, when the first simplified models were developed [12]. The two node model applied thermodynamics principles to energy exchanges between the human body and its thermal environment [13] for the first time during the 1930s, but it is only from the 1960s onwards that researchers were able to analyze the main climatic parameters connected to the perception of thermal comfort (e.g., air temperature, radiant temperature, air humidity, air flow velocity) when the first climate chambers were made available [14]. The cornerstone studies of Givoni [15] and Fanger [16] led in the following years to the identification of new parameters that are currently considered essential elements in the contemporary assessment of thermal comfort. The advances in the physics of heat exchange knowledge gained during the 1980s and the increasing availability of computer tools to support the research activity allowed relevant progress on the understanding of the human thermal environment [17–22] and the formulation of indexes based on body heat exchange [11].

In order to model human thermal comfort in outdoor environments, solar radiation was first added to the set of climatic variables in use for indoor spaces [23,24]. Olgyay assumed that solar radiation must be combined to the effects of other climatic elements, to draft a "bioclimatic chart" for the outdoor conditions [23].

Further studies have shown that outdoor thermal comfort is a more complex notion and a multilayered condition, which is very difficult to properly describe as a whole by considering biometeorological factors only [25]. Although the thermal state appears as very influential among the many factors shaping the quality of outdoor spaces, a wide range of additional social and physical aspects, however, were identified as relevant, especially those linked to behavioral variables [26].

Nonetheless, the issue remains open, especially regarding the assessment of the human variables influenced—including cultural, behavioral and psychological factors—on the perception of the environment's physical conditions [10].

The efforts spent in the last few decades to advance the understanding of these phenomena provide evidence of the topic's relevance, even if a comprehensive picture of available studies is still lacking, as well as a classification of the interconnected subjects and outcomes. A systemic overview of the available knowledge could be therefore a useful tool for identifying the different research trends and classifying their objectives, approaches, results and implications.

This paper reports the outcome of a literature review aimed at screening the available resources dealing with outdoor thermal comfort, in order to provide a state of the art that identifies the main topics focused on by the researchers, as well as the barriers in defining suitable indexes and approaches for thermal comfort assessment in outdoor environments.

1.2. Theoretical Background

The International Organization for Standardization (ISO) has released a series of international regulations for the evaluation of thermal comfort; ISO 13731:2001 defines physical quantities and provides a reference for terminology and symbols to adopt for standards on ergonomics of the thermal environment [27], while ISO 7726:1998 identifies the means and instruments for measuring the physical quantities involved [28]. ISO 7730:2005 provides an analytical determination and interpretation of thermal comfort using calculation of the Predictive Mean Vote (PMV) and Percentage of Person Dissatisfied (PPD) indexes and local collected data [29]. Although the model was developed for indoor comfort assessment, it may be adapted for outdoor spaces by adding the radiative exchange values [30].

This ISO Standard also includes annexes providing comprehensive databases for the metabolic rates of different human activities and the thermal insulation values of clothing ensembles. Moreover, ISO 7243:2017 enables the estimation of the worker heat stress by Wet Bulb Globe Temperature Index (WBGT) [31], which can be applied for both indoor and outdoor work environments.

In addition to ISO standards, other regulations such as American National Standards Institute (ANSI) / American Society of Heating, Refrigerating and Air-Conditioning Engineers (ASHRAE) Standard 55-2017 and European Standards (EN) 15251-2017 specify calculation and evaluation methods for thermal comfort. However, they are mostly addressed to indoor environments, and focused on those parameters affecting the energy performance of buildings [32].

The awareness of the topic's broad latitude has prompted some authors to focus on the possible gaps in criteria and assumptions currently in use to map the factors influencing the perceived outdoor comfort and wellbeing sensation, starting from the definition of thermal comfort itself.

ASHRAE defines thermal comfort as a state of mind that expresses satisfaction with the thermal surroundings [33], which means that human thermal comfort refers to a subjective sensation, different from one subject to another [34]. Some studies argue that this definition may appear rather vague [10]: it does not specify what that state of mind is (in terms of perception, feeling, etc.) and it does not provide any indication of how to relate this mental state into something that can be measured, nor which variables could be involved [35]. Thus, this is still an open issue from different points of view, although the definition is intended to be the most general as possible, to provide a common understanding, thus leaving each study the responsibility to state the assumptions (and limitations) in their own premises.

Additionally, when the assessment of this mental state has to be investigated, referring to the outdoor environment, the relationship between the human body and a large set of spatial and temporal variables must be also considered. Theoretically, this gap could be filled by adapting for the outdoor comfort assessment the same methodologies and indexes developed to evaluate indoor comfort. However, several authors discussed this position as unsuitable, arguing that the theoretical models developed for describing thermoregulation functions within the indoor environment are not adequate to feature the outdoor thermal comfort conditions [34,36]. This is mainly because of the outdoor environment's greater complexity, and its temporal and spatial variability [34]. Thus, the need is acknowledged for empirical data from field surveys on the subjective human perception of outdoor wellbeing, which should enable investigation of thermal comfort in open spaces from a broader and more realistic perspective [34].

In order to make the reading easier, Table 1 provides the nomenclature of the main terms and acronyms reported in the paper, as well as Table 2, which summarizes the main thermal comfort indexes.

Table 1. Nomenclature of main terms used.

Abbreviation	Definition
UHI	Urban Heat Island
IPCC	Intergovernmental Panel on Climate Change
NOAA	National Oceanic and Atmospheric Administration

Table 2. Nomenclature of the main thermal comfort indexes cited.

Abbreviation	Index	Unit
ASV	Actual Sensation Vote	-
AT	Apparent Temperature	°C
COMFA	COMfort FormulA	$W{\cdot}m^{-2}$
DI	Discomfort Index	°C
ESI	Environmental Stress Index and	°C
ET	Effective Temperature	°C
ETU	Universal Effective Temperature	°C
H	Humidex	°C
HI	Heat Index	°C
ITS	Index of Thermal Stress	W
MENEX	Man ENvironmental Heat EXchange model	$W{\cdot}m^{-2}$
OUT_SET*	Outdoor Effective Temperature	°C
PE	Cooling Power Index	$kcal{\cdot}m^{-2}{\cdot}h$
PET	Physiologically Equivalent Temperature	°C
PMV	Predicted Mean Vote	-
PSI	Physiological Strain Index	°C
PT	Perceived Temperature	°C
RSI	Relative Strain Index	-
SET	New Standard Effective Temperature	°C
TS	Thermal Sensation	-
TSV	Thermal Sensation Vote	-
UTCI	Universal Thermal Climate Index	°C
WBGT	Wet Bulb Globe Temperature Index	°C
WCI	Wind Chill Index	°C

2. Methodology

Although overall human comfort involves several environmental agents acting simultaneously, including air quality and thermal, acoustic, and lighting factors [34,37], this study was limited to the thermal wellbeing, since it plays a crucial role in affecting the comfort perception in outdoor environments.

The review was based on a systematic search for peer-reviewed papers published within the last twenty years. The aim of the review was to identify the most relevant trends in studies dealing with outdoor human thermal comfort. Five main search engines were used: Science Direct, Google Scholar, Scopus, Web of Science (WOS) and Researchgate. The following keywords were used for the preliminary retrieval of papers from the sources:

- outdoor thermal comfort;
- thermal perception;
- thermal wellbeing;
- human thermal comfort;
- human thermal index;
- outdoor thermal comfort approaches;
- thermal comfort assessment.

This allowed a first selection based on the paper title and abstract. Additionally, papers referenced within the selected articles were considered as secondary sources, thus embedded in the second step of the review process.

Review Process and Outcomes

The outcome of the first search round provided more than 25,000 results, of which 1059 were retrieved from the Science Direct database, 16,600 from the Google Scholar search engine, 710 from the

Scopus database, 594 from Web of Science and 6160 from Researchgate. Duplications were deleted in a second step of the process and the results were also filtered using a combination of the proposed keywords. The results distribution from Science Direct, reported in Figure 1, indicates a growing interest in the second decade that can be certainly associated to new drivers, such as the effects of climate change and related heat waves, but also to the development of web-based solutions to share knowledge and studies that facilitated the communication and exchange among the scientific community. Furthermore, it must be noted that the increasing demand for scientific publications on the topic within the academic circuits, for both research purposes and career advancement, may have influenced the numeric growth of studies, as well as their availability in scientific journals.

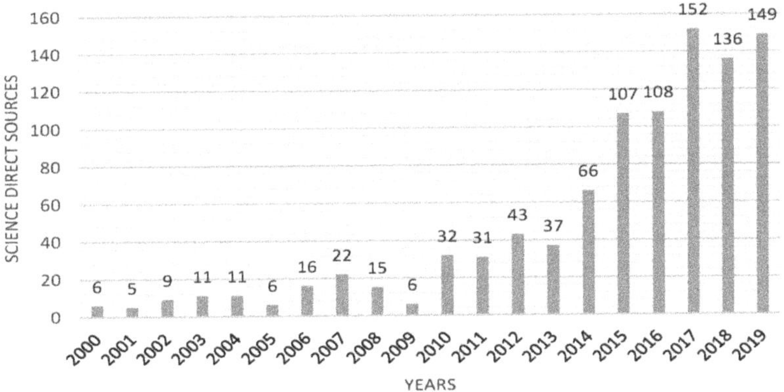

Figure 1. Results retrieved from the ScienceDirect search engine (keys: publication after 1999; text string "outdoor thermal comfort" in paper abstract and/or title).

The selection results were then refined according to the predefined set of keywords; a total of 855 sources were found at the end of the filtering process, including journal articles, book chapters, reviews and peer-reviewed conference papers. The sources matching with three or more keywords (out of six) were considered as having highly relevant contents. They were then shortlisted and their full text downloaded; 146 significant outputs were identified by selecting those focusing on the relationships between outdoor thermal comfort and microclimatic variables.

The sources referenced by the selected papers were also explored, thus increasing to 236 the final number of the surveyed articles. Figure 2 displays the incidence per year (Figure 2a) and the breakdown by issues addressed (Figure 2b).

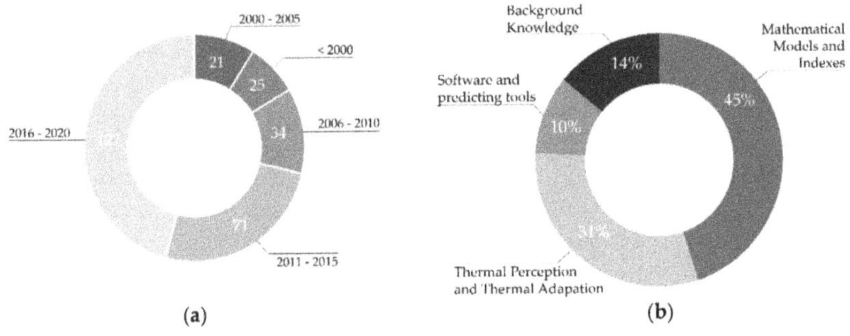

Figure 2. Distribution by year (**a**) and by topic (**b**) of the final 236 sources considered for the review purposes.

The analysis of the 236 final resources, as shown in Figure 2b, pointed out that studies on outdoor comfort could be divided into four main groups:

- The first group [1–5,11–13,16,19–24,26–29,31–33,38–47] includes papers that provide theoretical background elements as well as definitions, terminology and regulations (14% of items);
- The second group [8,14,15,17,18,22,30,48–149] collects mathematical models and indexes for the definition of new approaches in outdoor comfort assessment (45% of selected items);
- The third group [6,7,9,10,25,34–37,150–213] includes investigations of the physical, physiological and psychological human adaptability as a key for understanding thermal perception in outdoor environments (31% of items);
- The fourth group [214–236] deals with the use of software and forecasting tools to virtually reproduce complex environmental contexts, in particular with the aim of supporting the designer in understanding the effects of changes in the climatic factors that affect people's external comfort (10% of articles).

Accordingly, this review adopts the same structure, reporting the findings grouped in three sections, corresponding to the main topics to which scientific efforts were devoted toward identifying both effective methodologies and main limitations for the full comprehension of the subject. In addition, a brief introduction to standards and regulations is provided in the theoretical background paragraph of the introduction.

From the literature analysis, field studies on outdoor thermal comfort were carried out around the world in the last 20 years. Givoni et al. discussed methodological issues and deepens problems in outdoor comfort research based in Japan and Israel [164], as well as in China [74], where other studies were also conducted [51,54,57]. Other studies were performed with reference to the following geographical areas: Canada [207], Argentina [138], Sweden [147,175,218], Portugal [131], United Kingdom [187,209], Italy [214], Morocco [132], Emirates [227], Egypt [118], Malaysia [192], Bangladesh [141], Australia [142], New Zealand [208]. Among the largest research projects, the most extensive was RUROS: Rediscovering the Urban Realm and Open Spaces [37], which included field surveys carried out in seven European cities: Athens, Thessaloniki, Milan, Fribourg, Kassel, Cambridge, Sheffield.

3. Results

3.1. Mathematical Models and Indexes

Several complex thermal indexes were developed to date, describing and quantifying the thermal environment of humans and the energy fluxes between the human body and the surrounding environment. De Freitas and Grigorieva [79,80] carried out a three-stage study, providing a comprehensive register of 165 indexes suitable for the purpose, which were subsequently grouped and classified. Thus, they observed that indexes are almost designed for a specific application, so the choice depends on the context in which the index is used, as well as on the availability of the data needed to quantify it. They also found that the best performing indexes are those based on the body/atmosphere energy balance, which, however, are those needing more complicated calculation routines and more detailed input data. What emerges as an additional and more serious drawback is that body–atmosphere energy balance indexes are often based on numerical models that were not validated. This leads to the conclusions that there is not an overall best index [79] and the use of a standardized human body could introduce errors, since the characteristics of the human body vary individually [80].

A review about models and standards is provided by Coccolo et al. [157] who analyzed a number of outdoor human comfort models and the related physical variables as well as the applicability with reference to the climate, with reference to the research goals, dividing the models in the following three groups:

- Indexes based on the human's energy balance, which show the interrelation between metabolic activities, clothing and environmental parameters, and humans thermal perception. They include COMFA (COMfort FormulA), ETU (Universal Effective Temperature), ITS, MENEX, PET, PMV, PT, OUT_SET*, SET, UTCI;
- Empirical indexes, which are expressed as linear regressions based on field studies (monitoring and surveys) defining the human comfort for a specific climate or location that are set and validated for it. They consider Actual Sensation Vote (ASV), Thermal Sensation (TS), Thermal Sensation Vote (TSV);
- Indexes based on linear equations defining the comfort as function of the thermal environment, by focusing on air temperature, wind speed and relative humidity parameters, but neglecting the microclimate and human behaviour. Among these are Apparent Temperature (AT), Discomfort Index (DI), Environmental Stress Index and (ESI) and Physiological Strain Index (PSI), Effective Temperature (ET), Humidex (H), Heat Index (HI), Cooling Power Index (PE), Relative Strain Index (RSI), Wet Bulb Globe Temperature Index (WBGT), Wind Chill Index (WCI).

A selection of case studies about the different models since 2000 is organized in a graph according to the Köppen climate classification by Coccolo et al [157]. Temperate climate emerges as the most studied condition, followed by those referring to arid, cold and tropical climates, while very few studies were performed for polar environments. The outcomes of this research leads to the following conclusions:

- Thermal indexes based on energy balance enable quantification of thermal sensation from the general climate to the urban microclimate, considering the human variables.
- Empirical indexes reliably describe the thermal perception of humans and the environmental factors affecting their thermal behavior.
- Indexes based on linear equations do not allow for more comprehensive microclimate analysis, even if they can be useful for meteorological forecasting or for mapping of thermal comfort trends over the time [86].

From the literature analysis, it emerges that PMV and PET are the most widely used indexes [49,50,59,61,62,67,71,75,77,80,85,86,88,89,92,93,95,98,109,110,112,114,118,128,131,133,138,141, 142,148,151,154]. However, PMV was elaborated for indoor environments by observing people sitting in climate chambers; it does not consider the dynamic adaptive response of the human body and instead references a punctual static situation. Thus, its use in outdoor environments may often give misleading results. Cheng et al. [74] clearly demonstrate that PMV generally overestimates the thermal sensation in summer and underestimates in winter.

Developed for the outdoor environment, PET is a procedure often adopted, with results that are well correlated with onsite monitoring and questionnaires [49,77,89,114,131,133,141,142,148]. The PET index can be used as an alternative to PMV; however, the main drawback is that it can underestimate the effect of latent heat fluxes and overestimate radiant heat flows [74]. By comparing the outdoor thermal environment and the thermal sensation of pedestrians in two different Chinese cities, and through the thermal unacceptability percentage and PET indexes, Yang et al. [54] observed that the thermal unacceptability expressed by people was different according to the city, despite the similar outdoor thermal conditions.

In arid climates, ITS results show high correlation with field studies [196,200], while SET [52,142,191] and OUT SET* [133,148] appear to be more applied in and reliable for temperate climates. Being more scientifically updated, UTCI is instead the only index that was applied to all climates [59,84,99], ensuring a good matching between onsite measures and simulations [70,133,157]. PE and WCI, describing thermal sensations from comfort to extreme cold stress are preferable in cold climates. ITS, H, HI and WGBT make available detailed thermal scales for hot sensations, often neglecting the cold ones [157].

From the research carried out so far in order to understand the primary models used, it seems that there are apparently a large number of indexes available for the assessment of outdoor comfort, despite each of them presenting some drawback with different levels of error or approximation.

Although energy balance based indexes are widely used, the main limitation lies in the steady-state condition that does not fully reflect people rarely experience thermal equilibrium in outdoor environments [100].

A positive note is that the scientific community seems to be aware of this limit and much effort was spent to understand the human adaptation capacity as well [25].

3.2. Human Thermal Perception and Thermal Adaptation

The term adaptation can be broadly defined as the gradual decrease of the organism's response under an iterated exposure to a stimulus, thanks to the effects of all the actions deployed by the subject to make it better suited to survive in such an environment. In the context of thermal comfort, this can involve all the processes that people perform to reduce the gap between the environmental conditions and their requirements [194]. In other words, whenever metabolic activity or environmental conditions change, the human body tends to adapt itself to those changes. According to Nikolopoulou and Steemers [194], the adaptation basically occurs in three different ways:

- Physical adaptation—namely, the changes that a person makes in order to adjust oneself to the environment (such as altering clothing layers, posture and position, or drinking) or to conform the environment to his needs;
- Physiological adaptation—also called physiological acclimatization, which implies changes in the physiological response mechanisms resulting from repeated exposure to a stimulus;
- Psychological adaptation—which involves the different ways that individual people perceive the environment, being the human response is not only direct related to the physical stimulus magnitude, but also to the information that people have regarding that situation. The familiarity with that climate, the individual expectations, experiences, time of exposure and alleged control power on the situation, significantly influence the perception of environmental stimuli. Cultural factors and personal attitudes also affect the thermal perception, thus underlining the need to connect the thermal comfort indexes to the emotional feeling individually established with the environment [175].

A large number of studies were done in the last twenty years that aimed to incorporate the human dimensions into comfort assessment methods, performed both through climate chambers and by direct field surveys. Some of these studies investigated the possible adaptation from a thermophysiological perspective [95,174], others focused on the parameters that determine the human perception of comfort [81,142,194,208].

Some links between human thermoregulation mechanisms and thermal environment conditions were established by running tests in climate chambers [149,195]. However, whether these results can be transferred to the behavior of people in external environments is still an open question, since all the aspects that influence adaptation actions in real contexts are highly complex to reproduce (providing wide temperature ranges, checking for human physical and behavioral changes, measuring temperatures of people' skin and core to evaluate thermoregulation features etc.).

Thus, the assessment of the human thermal sensation must consider the environmental stimuli, which are dynamic and perceived subjectively. The stimuli are dynamic due to the human adaptation to external climatic conditions, which is a progressive process influenced by various adaptive factors. It is subjectively perceived, since the human perception of thermal comfort is not always nor univocally dependent only on objective biometeorological conditions. This means that the individual attitude towards outdoor space is not only determined by the state of the body, but also by the state of the mind. This suggests that the ideal framework for thermal comfort assessment should work on at least four levels: physical, physiological, psychological and social/behavioral [7].

Therefore, by considering simultaneously all the factors (whether objective or subjective) influencing human thermal perception, it is possible to obtain an evaluation of outdoor comfort as coincident as possible with reality. Each of these factors can be estimated or calculated through different approaches (measurements, modeling, field interviews and observations). Thus, working on the four levels of evaluation of human outdoor thermal comfort, it should make it possible to connect the external microclimatic conditions with the perception of people who use a certain space at a specific temporal moment.

In other words, this framework should allow a linkage of "climatic knowledge" with "human knowledge" [7].

"Neuroarchitecture" seems to be opening a new research field that is able to drive some advances in this direction, combining neuroscience and architecture to better understand how space is perceived by the human brain [159,160,165]. The outcome from neuroarchitecture studies could thus enhance the effectiveness of the design of the built environment, providing a better knowledge of the relationship between the humans and their spatial wellbeing [157].

Coburn et al. [202] investigated how neuroarchitecture could mature into an experimental science by outlining the related challenges ahead and identifying the priority need for a specific framework to guide research. To date, however, relatively little work has been done on architecture neuroscience, and further studies are needed.

Therefore, the final suggestion is to apply a multidisciplinary approach, including both studies on physical phenomena and human psychology. For this reason, it is highly recommended that shared principles and definitions be included in the common framework.

3.3. Software and Predictive Tools

Givoni et al. [164] addressed the need for prediction tools in order to support designers in understanding the effect of a change in a climatic element that influences people's outdoor comfort. In fact, the availability of simulation and scenario-testing tools within an assessment framework is crucial, as they provide a platform for both the integration of knowledge from various perspectives and the comparisons of different design options.

Currently, tools for simulating virtual scenarios are becoming increasingly available and updated, allowing reproduction of even complex environmental contexts. The literature review performed identified the most used software: ENVI-Met, RayMan, SOLWEIG and the UTCI calculator (Figure 3). While ENVI-Met is based on computational fluid dynamics (CFD) and thermodynamics, RayMan and SOLWEIG are basically 3D radiation models.

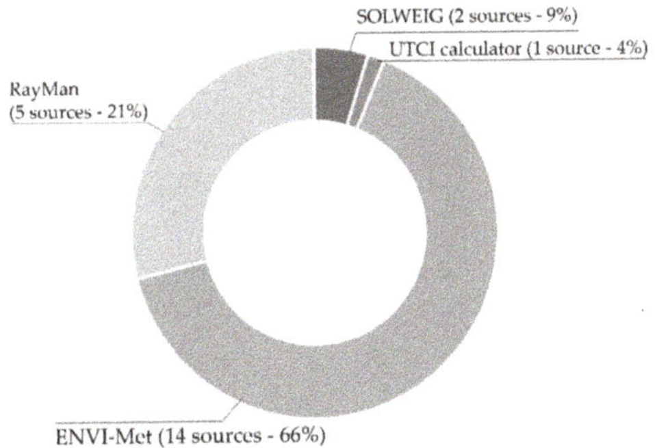

Figure 3. The most used software and tools outlined from the literature review.

ENVI-Met is a tool to simulate outdoor space microclimate by quantifying energy and mass exchanges, wind turbulence, vegetation effect on the outdoor conditions, bioclimatology data and pollution scattering. It is based on four interrelated systems: soil, vegetation, atmosphere and buildings.

Outdoor microclimate is described by air temperature, Mean Radiant Temperature (MRT), wind speed and direction, short- and long-wave radiation from a single building to an entire city [236]. It was used by Acero et al. to evaluate the differences in thermal comfort comparison models and onsite measures in four different locations [58]. The study points out that some deviations may occur in the ENVI-Met output; however, the tool provides useful and quite reliable outcomes (e.g., comparison of urban planning scenarios during typical meteorological conditions). Nevertheless, limitations must be clearly outlined in order to avoid misleading results. ENVI-Met is often used while interacting with other tools and plugins. Through the postprocessing tool called BioMet, it is possible to determine thermal comfort according to PMV, PET, UTCI [215] and MRT. Additionally, thanks to a generative algorithm called ENVI-BUG Software, it is possible to combine ENVI-Met, Rhinoceros, Grasshopper and LadyBug. Fabbri et al. effectively adapted it to obtain a 3D output, achieving a simplified method for displaying results and making them easier to read for nonexpert users [230].

Developed at the University of Freiburg, RayMan is a diagnostic microscale radiation model able to calculate radiation fluxes and thermophysiologically indexes, such as PMV, PET, SET* [222], UTCI, PT [220] and MRT. It is mainly used to compare the effect of multiple planning scenarios in different situations from micro to regional scales [220]. It allows the use of several input data such fish-eye pictures or obstacle files to obtain additional outputs like shade and sunshine duration, as well as the possibility to run long-term data sets. The major drawback is that the model cannot calculate air temperature, air humidity and calculate or adjust wind speed. These gaps are often bridged by preparing the data in the input files or running simulations with different wind speeds [222].

SOLWEIG (SOlar and LongWave Environmental Irradiance Geometry) enables quantification of PET, UTCI and MRT within complex urban settings as described by Lindberg et al. [218]. It applies the theory of radiative fluxes and mean radiant temperature, the main limitation is that it takes only building geometry into account, while vegetation is not considered when mean radiant temperature is calculated.

The UTCI calculator allows determination of a pedestrian's thermal comfort according to the Universal Thermal Climate Index [117]. Abdel-Ghany et al. demonstrated its application in combination with RayMan to evaluate UTCI index in arid climatic conditions [128], concluding that the model can be used successfully in arid environments to evaluate the thermal sensation, with the heat stress outcome very close to the PET index.

These environmental modeling tools can provide a better understanding of climatic conditions and a mean to effectively assess human thermal comfort outdoors, helping town planners and decision makers to compare and test several design alternatives in terms of attractiveness and effectiveness [232]. In addition, the development of such tools and software can solve the limitation of the different methodologies used in research.

4. Discussion

The literature review in this paper lists the main available resources dealing with outdoor thermal comfort issues. Its aims to provide a state-of-the-art review to identify the main topics on which current research focuses, and what the main barriers are that limit the identification of successful indexes for the assessment of thermal comfort in the outdoor environment, especially in urban contexts [150,170].

The main outcome is that thermal comfort in the outdoor environment is a complex issue with multiple layers and that the human state of the mind plays a key role in influencing peoples' perception on space and its utilization.

The literature review performed points out that many indexes and approaches have been developed, however they are specifically addressed to meet particular contexts with relation to specific variables. A relevant drawback deals with the limited consideration of the dynamic adaptive response of the human body, since the most frequently adopted indexes focus on the energy balance between the human body and the environment. More recent indexes seem to pay more attention to human perception and behavior, but their application is still limited and therefore a discussion on the potential outcomes is still hard and may be somehow misleading. The main suggestion is to focus on a limited and possibly shared number of procedures that adequately consider the human thermal perception and thermal adaptation.

It must be noted, despite some attempts to integrate different disciplines on the same topic, that most studies are organized by adopting a silos approach.

Moreover, many studies reveal that different people experience the environment in a different way. The human response to a physical stimulus is not in direct relationship to its intensity, but depends on the "information" that people have for a particular situation, and on the associated psychological factors influencing the thermal perception of a space and the changes occurring in it [194]. If physiological acclimatization is not sufficient to meet a comfort status, physical adaptation will be introduced to adjust oneself to the environment or alter the environment to his needs (such as altering clothing levels, modifying posture and position, or even changing metabolic heat with the consumption of hot or cool drinks).

Since the agents acting on this mechanism belong to at least four different but interconnected patterns (physical, psychological, physiological and social/behavioral), the ideal framework for thermal comfort assessment should work on all of them.

No effective methods are available today that include the human dependent factors within the outdoor thermal comfort models. However, the scientific community seems to become increasingly aware of the importance of the psychological and social/behavioral factors, as indicated by the number of recent studies exploring these fields. Integrating the physical energy balance and the human variables would allow for creation of more complete and thus more effective models, drastically improving the reliability of their results.

Despite the complexity of the above interrelations, these topics should be approached at design level. More effective simulation and modeling tools could be developed within that framework, providing designers and decision makers with a means to better achieve the thermal outdoor comfort target by integrating the factors related to climatic conditions and those belonging to the people's sensitivity to environmental stimuli; these tools could be wisely adopted in urban design. In this way, shopkeepers could be the first group to realize the benefit of such cool oases in a hot environment, and finally it would be possible to extend the positive impacts not only to the environmental domain of cities but also to the economic domain.

The highly advisable trend sketched by this scenario cannot hide the fact that shortcomings are still evident in research, concerning both the tools and the goals. Concerning tools, the systematization of all knowledge belonging to physical, physiological and psychological studies seems to be the only way forward today for the identification of an effective and shared method to evaluate the achievement of thermal comfort in outdoor spaces. Regarding goals, deeper interdisciplinary studies are needed to support with evidence the assumption that the more intense use of outdoor space will benefit the economic and social life of the city.

5. Conclusions

The literature review performed points out that many different interrelated issues drive the research on outdoor thermal comfort. The topic emerges as worthy to be further investigated, especially in relation to social and behavioral implications, requiring to possibly adopt a transdisciplinary approach. Despite the effort spent to create accurate models and software to properly assess outdoor human comfort, the evocated state of mind of the final user still remains at the core of the uncertain process. The field studies, including surveys and tests with different categories of users, highlight the need to refine the available indexes to better reflect the real perception of the human body in different conditions.

Therefore, each user has to clearly understand the basic equations of the software or tool chosen, in order to select the one that best suits the needs for the research purpose and the application context. This often generates some uncertainty that makes a comparative approach among the different outcomes more difficult.

Some relevant shortcomings can be currently detected and listed as follows, with the aim to address and prioritize the research efforts for further improving an understanding of outdoor human comfort.

Assuming that an effective assessment requires consideration of physical, physiological, psychological and behavioral levels, the detected lack of shared principles and definitions within a common framework reduces the possibility to take into account the multiple and interconnected nature of phenomena. Accordingly, the definition of a comprehensive and stable framework represents a top priority:

- The availability of several models and methods is on the one hand a true sign of interest in assuming outdoor thermal comfort as a relevant field of research, especially when connected to climate change effects (UHI, heat waves etc.), but each methodology has its own limitations and the differences make it harder to compare the outcomes. The development or the refinement of tools and software able to solve these gaps may certainly support the research activity in the future.
- The gap between models and real perception in experimental studies suggests that human sensation and behavior are central and crucial elements to further improving the quality of research.
- Understanding the outcomes is another relevant issue dealing with a proper communication of the possible social implications to nonexperts. Even if this is not a top priority, it will certainly contribute by increasing the attention towards quality of life in outdoor urban spaces.

The use of reliable predictive and simulation tools will support decision makers, designers and planners to better realize the potential impacts of their decision and strategies when transforming the built environment. Finally, a multidisciplinary approach, including studies on physical phenomena, human psychology and architecture, is highly recommended, assuming people's wellbeing is the ultimate goal.

Author Contributions: Conceptualization, E.A. and J.G.; methodology, V.V., E.A. and J.G.; formal analysis, V.V.; investigation, V.V.; resources, M.D.G. and V.V.; writing—original draft preparation, V.V.; writing—review and editing, E.A. and J.G.; visualization, V.V.; supervision, E.A. and J.G. All authors have read and agreed to the published version of the manuscript.

Funding: This research received no external funding.

Conflicts of Interest: The authors declare no conflict of interest.

References

1. Population Reference Bureau 2008 WORLD POPULATION Data Sheet. *Popul. Bull.* **2008**, 10–11.
2. Akpınar, M.; Sevin, S. *Reducing Urban Heat Islands by Developing Cool Pavements*; Springer Science and Business Media LLC: Cham, Germany, 2018; pp. 43–50.
3. Hoegh-Guldberg, O.D.; Jacob, D.; Taylor, M.; Bindi, M.; Brown, S.; Camilloni, I.; Diedhiou, A.; Djalante, R.; Ebi, K.L.; Engelbrecht, F.; et al. Global Warming of 1.5 °C. *Special Report* **2014**, 177–287.
4. Rebecca, L.; Dahlman, L. Climate Change: Global Temperature|NOAA Climate.gov. Available online: https://www.climate.gov/news-features/understanding-climate/climate-change-global-temperature (accessed on 27 March 2020).
5. For Some Urban Areas, A Warming Climate is only Half The Threat–Science Daily. Available online: https://www.sciencedaily.com/releases/2019/11/191114115946.htm (accessed on 27 March 2020).
6. Gehl, J. *Life between Buildings: Using Public Space*; Island Press: Washington, DC, USA, 2011.
7. Chen, L.; Ng, E. Outdoor thermal comfort and outdoor activities: A review of research in the past decade. *Cities* **2012**, *29*, 118–125. [CrossRef]
8. Maruani, T.; Amit-Cohen, I. Open space planning models: A review of approaches and methods. *Landsc. Urban Plan.* **2007**, *81*, 1–13. [CrossRef]
9. Nikolopoulou, M.; Baker, N.; Steemers, K. Thermal comfort in outdoor urban spaces: Understanding the human parameter. *Sol. Energy* **2001**, *70*, 227–235. [CrossRef]

10. Nikolopoulou, M. Outdoor thermal comfort. *Front. Biosci. Sch.* **2011**, *3*, 1552–1568. [CrossRef]

11. Fabbri, K. *Indoor Thermal Comfort Perception*; Springer International Publishing: Cham, Switzerland, 2015.

12. Gagge, A.P. An effective temperature scale based on a simple model of human physiological regulatory response. *Ashrae Trans.* **1971**, *77*, 247–262.

13. Gagge, A.P. The linearity criterion as applied to partitional calorimetry. *Am. J. Physiol. Content* **1936**, *116*, 656–668. [CrossRef]

14. Honjo, T. Thermal Comfort in Outdoor Environment. *Global Environ. Res.* **2009**, *13*, 43–47.

15. Givoni, B. *Estimation of the Effects of Climate on Man: Development of a New Thermal Index*; Technion: Haifa, Israel, 1963.

16. Fanger, P. *Thermal Comfort*; McGraw-Hill: New York, NY, USA, 1972.

17. Gagge, A.P.; Fobelets, A.P.; Berglund, L.G. Standard Predictive Index of Human Response to The Thermal Environment. In *Proceedings of the ASHRAE Transactions*; ASHRAE: Portland, OR, USA, 1986; Volume 92, pp. 709–731.

18. Höppe, P. The physiological equivalent temperature—A universal index for the biometeorological assessment of the thermal environment. *Int. J. Biometeorol.* **1999**, *43*, 71–75. [CrossRef]

19. Mayer, H.; Höppe, P. Thermal comfort of man in different urban environments. *Theor. Appl. Clim.* **1987**, *38*, 43–49. [CrossRef]

20. Steadman, R.G. Measurements of dry atmospheric cooling in subfreezing temperatures. *Proc. Am. Philos. Soc.* **1945**, *89*, 177–199.

21. Steadman, R.G. Indices of windchill of clothed persons. *J. Appl. Meteorol.* **1971**, *10*, 674–683. [CrossRef]

22. Steadman, R.G. A Universal Scale of Apparent Temperature. *J. Clim. Appl. Meteorol.* **1984**, *23*, 1674–1687. [CrossRef]

23. Olgyay, V. *Design with Climate: Bioclimatic Approach to Architectural Regionalism-New and Expanded Edition*; Princeton Univerity Press: Princeton, NJ, USA, 2015.

24. Penwarden, A. Acceptable wind speeds in towns. *Build. Sci.* **1973**, *8*, 259–267. [CrossRef]

25. Perera, K.; Schnabel, M.A.; Donn, M.; Maddewithana, H. Addressing Human Thermal Adaptation in Outdoor Comfort Research—A Literature Review. In Proceedings of the Making built environments responsive, 8th International Conference of Faculty of Architecture Research Unit (FARU), University of Moratuwa, Sri Lanka, 11–12 December 2015; pp. 477–490.

26. Parsons, K. *Human Thermal Environments: The Effects of Hot, Moderate, and Cold Environments on Human Health, Comfort, and Performance*, 3rd ed.; CRC Press: Boca Raton, FL, USA, 2014; ISBN 9781466595996.

27. ISO 13731:2001-Ergonomics of the Thermal Environment—Vocabulary and Symbols. Available online: https://www.iso.org/standard/22450.html (accessed on 19 February 2020).

28. ISO 7726:1998-Ergonomics of the Thermal Environment—Instruments for Measuring Physical Quantities. Available online: https://www.iso.org/standard/14562.html (accessed on 19 February 2020).

29. ISO 7730:2005-Ergonomics of the thermal environment—Analytical Determination and Interpretation of Thermal Comfort using Calculation of the PMV and PPD Indices and Local Thermal Comfort Criteria. Available online: https://www.iso.org/standard/39155.html (accessed on 29 January 2020).

30. Jendritzky, G.; Nübler, W. A model analysing the urban thermal environment in physiologically significant terms. *Theor. Appl. Clim.* **1981**, *29*, 313–326. [CrossRef]

31. ISO 7243:2017-Ergonomics of the Thermal Environment—Assessment of Heat Stress Using the WBGT (Wet Bulb Globe Temperature) Index. Available online: https://www.iso.org/standard/67188.html (accessed on 19 February 2020).

32. EN 15251:2007–Policies–IEA. Available online: https://www.iea.org/policies/7029-en-152512007 (accessed on 19 February 2020).

33. *American Society of Heating Refrigeration and Air-Conditioning Engineers ASHRAE Handbook Fundamentals*; ASHRAE: Atlanta, GA, USA, 1989.

34. Nikolopoulou, M.; Lykoudis, S. Thermal comfort in outdoor urban spaces: Analysis across different European countries. *Build. Environ.* **2006**, *41*, 1455–1470. [CrossRef]

35. Heijs, W.J.M. The dependent variable in thermal comfort research some psychological considerations. In *Proceedings of the Thermal Comfortpast Present and Future: Proceedings of the Conference of June 1993*; Oseland, N., Humphreys, M., Eds.; Watford Building Research Establishment: Garston, UK, 1994; pp. 40–51.

36. Shooshtarian, S. Theoretical dimension of outdoor thermal comfort research. *Sustain. Cities Soc.* **2019**, *47*, 101495. [CrossRef]
37. RUROS Database: Field Survey Results. Available online: http://alpha.cres.gr/ruros/database.htm (accessed on 16 February 2020).
38. US Department of Commerce National Oceanic and Atmospheric Administration. Available online: https://www.noaa.gov/ (accessed on 30 March 2020).
39. ISO 11079:2007-Ergonomics of the Thermal Environment—Determination and Interpretation of Cold Stress When Using Required Clothing Insulation (IREQ) and Local Cooling Effects. Available online: https://www.iso.org/standard/38900.html (accessed on 29 January 2020).
40. ISO 7243:1989-Hot Environments—Estimation of the Heat Stress on Working Man, Based on the WBGT-Index (Wet Bulb Globe Temperature). Available online: https://www.iso.org/standard/13895.html (accessed on 29 January 2020).
41. Cheung, P.K.; Jim, C. Determination and application of outdoor thermal benchmarks. *Build. Environ.* **2017**, *123*, 333–350. [CrossRef]
42. Gagge, A.P.; Stolwijk, J.A.J.; Nishi, Y. An Effective Temperature Scale Based on a Simple Model of Human Physiological Regulatiry Response. *Memoirs Faculty Eng. Hokkaido Univ.* **1971**, *13*, 21–36.
43. Chun, C.; Kwok, A.; Tamura, A. Thermal comfort in transitional spaces-basic concepts: Literature review and trial measurement. *Build. Environ.* **2004**, *39*, 1187–1192. [CrossRef]
44. Thom, E.C. The Discomfort Index. *Weatherwise* **1959**, *12*, 57–61. [CrossRef]
45. EPA Heat Island Compendium. *US Environ. Prot. Agency* **2008**, 2–3.
46. Varquez, A.C.G.; Kanda, M. Global urban climatology: A meta-analysis of air temperature trends (1960–2009). *npj Clim. Atmospheric Sci.* **2018**, *1*, 32. [CrossRef]
47. Wu, H.; Wu, Y.; Sun, X.; Liu, J. Combined effects of acoustic, thermal, and illumination on human perception and performance: A review. *Build. Environ.* **2020**, *169*, 106593. [CrossRef]
48. Acero, J.A.; Arrizabalaga, J.; Kupski, S.; Katzschner, L. Urban heat island in a coastal urban area in northern Spain. *Theor. Appl. Clim.* **2012**, *113*, 137–154. [CrossRef]
49. Xi, T.; Li, Q.; Mochida, A.; Meng, Q. Study on the outdoor thermal environment and thermal comfort around campus clusters in subtropical urban areas. *Build. Environ.* **2012**, *52*, 162–170. [CrossRef]
50. Xie, Y.; Huang, T.; Li, J.; Liu, J.; Niu, J.; Mak, C.M.; Lin, Z. Evaluation of a multi-nodal thermal regulation model for assessment of outdoor thermal comfort: Sensitivity to wind speed and solar radiation. *Build. Environ.* **2018**, *132*, 45–56. [CrossRef]
51. Xie, Y.; Liu, J.; Huang, T.; Li, J.; Niu, J.; Mak, C.M.; Lee, T. cheung Outdoor thermal sensation and logistic regression analysis of comfort range of meteorological parameters in Hong Kong. *Build. Environ.* **2019**, *155*, 175–186. [CrossRef]
52. Yahia, M.; Johansson, E. Evaluating the behaviour of different thermal indices by investigating various outdoor urban environments in the hot dry city of Damascus, Syria. *Int. J. Biometeorol.* **2012**, *57*, 615–630. [CrossRef]
53. Yang, W.; Wong, N.H.; Jusuf, S.K. Thermal comfort in outdoor urban spaces in Singapore. *Build. Environ.* **2013**, *59*, 426–435. [CrossRef]
54. Yang, W.; Wong, N.H.; Zhang, G. A comparative analysis of human thermal conditions in outdoor urban spaces in the summer season in Singapore and Changsha, China. *Int. J. Biometeorol.* **2012**, *57*, 895–907. [CrossRef] [PubMed]
55. Yang, Y.; Gatto, E.; Gao, Z.; Buccolieri, R.; Morakinyo, T.E.; Lan, H. The "plant evaluation model" for the assessment of the impact of vegetation on outdoor microclimate in the urban environment. *Build. Environ.* **2019**, *159*, 106151. [CrossRef]
56. Yang, Y.; Zhou, D.; Wang, Y.; Ma, D.; Chen, W.; Xu, D.; Zhu, Z. Economical and outdoor thermal comfort analysis of greening in multistory residential areas in Xi'an. *Sustain. Cities Soc.* **2019**, *51*, 51. [CrossRef]
57. Zhao, L.; Zhou, X.; Li, L.; He, S.; Chen, R. Study on outdoor thermal comfort on a campus in a subtropical urban area in summer. *Sustain. Cities Soc.* **2016**, *22*, 164–170. [CrossRef]
58. Acero, J.A.; Herranz-Pascual, K. A comparison of thermal comfort conditions in four urban spaces by means of measurements and modelling techniques. *Build. Environ.* **2015**, *93*, 245–257. [CrossRef]
59. Achour-Younsi, S.; Kharrat, F. Influence of Urban Morphology on Outdoor Thermal Comfort in Summer: A Study in Tunis, Tunisia. *Mod. Environ. Sci. Eng.* **2016**, *2*, 251–256. [CrossRef]

60. Alfano, F.; Palella, B.I.; Riccio, G. Thermal environment assessment reliability using temperature–humidity indices. *Ind. Health* **2010**, *49*, 10081910028.

61. Ali-Toudert, F.; Djenane, M.; Bensalem, R.; Mayer, H. Outdoor thermal comfort in the old desert city of Beni-Isguen, Algeria. *Clim. Res.* **2005**, *28*, 243–256. [CrossRef]

62. Andreou, E. Thermal comfort in outdoor spaces and urban canyon microclimate. *Renew. Energy* **2013**, *55*, 182–188. [CrossRef]

63. Athamena, K.; Sini, J.-F.; Rosant, J.-M.; Guilhot, J. Numerical coupling model to compute the microclimate parameters inside a street canyon. *Sol. Energy* **2018**, *174*, 1237–1251. [CrossRef]

64. Balaras, C.; Tselepidaki, I.; Santamouris, M.; Asimakopoulos, D. Calculations and statistical analysis of the environmental cooling power index for Athens, Greece. *Energy Convers. Manag.* **1993**, *34*, 139–146. [CrossRef]

65. Balogun, I.A.; Daramola, M.T. The outdoor thermal comfort assessment of different urban configurations within Akure City, Nigeria. *Urban Clim.* **2019**, *29*, 100489. [CrossRef]

66. Beshir, M.; Ramsey, J.D. Heat stress indices: A review paper. *Int. J. Ind. Ergon.* **1988**, *3*, 89–102. [CrossRef]

67. Binarti, F.; Koerniawan, M.D.; Triyadi, S.; Utami, S.S.; Matzarakis, A. A review of outdoor thermal comfort indices and neutral ranges for hot-humid regions. *Urban Clim.* **2020**, *31*, 31. [CrossRef]

68. MENEX_2005 The Updated Version of Man-Environment Heat Exchange Model. Available online: http://www.igipz.pan.pl/tl_files/igipz/ZGiK/opracowania/indywidualne/blazejczyk/ (accessed on 30 March 2020).

69. Blazejczyk, K.; Bröde, P.; Fiala, D.; Havenith, G.; Holmér, I.; Jendritzky, G.; Kampmann, B.; Kunert, A. Principles of the New Universal Thermal Climate Index (UTCI) and its Application to Bioclimatic Research in European Scale. *Misc. Geogr.* **2010**, *14*, 91–102.

70. Blazejczyk, K.; Epstein, Y.; Jendritzky, G.; Staiger, H.; Tinz, B. Comparison of UTCI to selected thermal indices. *Int. J. Biometeorol.* **2011**, *56*, 515–535. [CrossRef]

71. Bouyer, J.; Vinet, J.; Delpech, P.; Carré, S. Thermal comfort assessment in semi-outdoor environments: Application to comfort study in stadia. *J. Wind. Eng. Ind. Aerodyn.* **2007**, *95*, 963–976. [CrossRef]

72. Bröde, P.; Fiala, D.; Blazejczyk, K.; Holmér, I.; Jendritzky, G.; Kampmann, B.; Tinz, B.; Havenith, G. Deriving the operational procedure for the Universal Thermal Climate Index (UTCI). *Int. J. Biometeorol.* **2011**, *56*, 481–494. [CrossRef]

73. Cheng, V.; Ng, E. Thermal Comfort in Urban Open Spaces for Hong Kong. *Arch. Sci. Rev.* **2006**, *49*, 236–242. [CrossRef]

74. Cheng, V.; Ng, E.; Chan, C.; Givoni, B. Outdoor thermal comfort study in a sub-tropical climate: A longitudinal study based in Hong Kong. *Int. J. Biometeorol.* **2011**, *56*, 43–56. [CrossRef] [PubMed]

75. Cheung, P.K.; Jim, C. Comparing the cooling effects of a tree and a concrete shelter using PET and UTCI. *Build. Environ.* **2018**, *130*, 49–61. [CrossRef]

76. Chow, W.T.L.; Akbar, S.N.; Assyakirin, B.A.; Heng, S.L.; Roth, M. Assessment of measured and perceived microclimates within a tropical urban forest. *Urban For. Urban Green.* **2016**, *16*, 62–75. [CrossRef]

77. Cohen, P.; Potchter, O.; Matzarakis, A.A. Human thermal perception of Coastal Mediterranean outdoor urban environments. *Appl. Geogr.* **2013**, *37*, 1–10. [CrossRef]

78. De Dear, R.; Pickup, J. An Outdoor Thermal Comfort index (OUT-SET*)-Part I-The Model and its Assumptions SAMBA IEQ Monitoring System View Project Natural Ventilation in Buildings View. In Proceedings of the 15th International Congress of Biometeorology and International Conference on Urban Climatology, Sydney, Australia, 8–12 November 1999.

79. De Freitas, C.R.; Grigorieva, E.A. A comprehensive catalogue and classification of human thermal climate indices. *Int. J. Biometeorol.* **2014**, *59*, 109–120. [CrossRef]

80. De Freitas, C.R.; Grigorieva, E.A. A comparison and appraisal of a comprehensive range of human thermal climate indices. *Int. J. Biometeorol.* **2016**, *61*, 487–512. [CrossRef] [PubMed]

81. Deb, C.; Ramachandraiah, A. A simple technique to classify urban locations with respect to human thermal comfort: Proposing the HXG scale. *Build. Environ.* **2011**, *46*, 1321–1328. [CrossRef]

82. Dimoudi, A.; Kantzioura, A.; Zoras, S.; Pallás, C.; Kosmopoulos, P. Investigation of urban microclimate parameters in an urban center. *Energy Build.* **2013**, *64*, 1–9. [CrossRef]

83. Dimoudi, A.; Zoras, S.; Kantzioura, A.; Stogiannou, X.; Kosmopoulos, P.; Pallás, C. Use of cool materials and other bioclimatic interventions in outdoor places in order to mitigate the urban heat island in a medium size city in Greece. *Sustain. Cities Soc.* **2014**, *13*, 89–96. [CrossRef]

84. Du, J.; Sun, C.; Xiao, Q.; Chen, X.; Liu, J. Field assessment of winter outdoor 3-D radiant environment and its impact on thermal comfort in a severely cold region. *Sci. Total. Environ.* **2020**, *709*, 136175. [CrossRef]

85. Ebrahimabadi, S. Outdoor Comfort in Cold Climates: Integrating Microclimate Factors in Urban Design. Ph.D. Thesis, Luleå Univ. Technol. 1 jan 1997, Luleå, Sweden, 2015; p. 195.

86. Emmanuel, R. Thermal comfort implications of urbanization in a warm-humid city: The Colombo Metropolitan Region (CMR), Sri Lanka. *Build. Environ.* **2005**, *40*, 1591–1601. [CrossRef]

87. Epstein, Y.; Moran, D.S. Thermal comfort and the heat stress indices. *Ind. Heal.* **2006**, *44*, 388–398. [CrossRef] [PubMed]

88. Evola, G.; Gagliano, A.; Fichera, A.; Marletta, L.; Martinico, F.; Nocera, F.; Pagano, A. UHI effects and strategies to improve outdoor thermal comfort in dense and old neighbourhoods. *Energy Procedia* **2017**, *134*, 692–701. [CrossRef]

89. Fang, Z.; Xu, X.; Zhou, X.; Deng, S.; Wu, H.; Liu, J.; Lin, Z. Investigation into the thermal comfort of university students conducting outdoor training. *Build. Environ.* **2019**, *149*, 26–38. [CrossRef]

90. Fiala, D.; Havenith, G.; Bröde, P.; Kampmann, B.; Jendritzky, G. UTCI-Fiala multi-node model of human heat transfer and temperature regulation. *Int. J. Biometeorol.* **2011**, *56*, 429–441. [CrossRef]

91. Foda, E.; Sirén, K. A new approach using the Pierce two-node model for different body parts. *Int. J. Biometeorol.* **2010**, *55*, 519–532. [CrossRef]

92. Golasi, I.; Salata, F.; Vollaro, E.D.L.; Coppi, M. Complying with the demand of standardization in outdoor thermal comfort: A first approach to the Global Outdoor Comfort Index (GOCI). *Build. Environ.* **2018**, *130*, 104–119. [CrossRef]

93. Golasi, I.; Salata, F.; Vollaro, E.D.L.; Coppi, M.; Vollaro, A.D.L. Thermal Perception in the Mediterranean Area: Comparing the Mediterranean Outdoor Comfort Index (MOCI) to Other Outdoor Thermal Comfort Indices. *Energies* **2016**, *9*, 550. [CrossRef]

94. Gonçalves, A.; Ribeiro, A.; Maia, F.; Nunes, L.; Feliciano, M. Influence of Green Spaces on Outdoors Thermal Comfort—Structured Experiment in a Mediterranean Climate. *Clim.* **2019**, *7*, 20. [CrossRef]

95. Gulyás, Á.; Unger, J.; Matzarakis, A. Assessment of the microclimatic and human comfort conditions in a complex urban environment: Modelling and measurements. *Build. Environ.* **2006**, *41*, 1713–1722. [CrossRef]

96. Guo, H.; Aviv, D.; Loyola, M.; Teitelbaum, E.; Houchois, N.; Meggers, F. On the understanding of the mean radiant temperature within both the indoor and outdoor environment, a critical review. *Renew. Sustain. Energy Rev.* **2020**, *117*, 109207. [CrossRef]

97. Hai, Y.; Feng, Q. Thermalscape of Ecological City and its Visualized Evaluation. *Energy Procedia* **2018**, *152*, 1139–1144. [CrossRef]

98. Hami, A.; Abdi, B.; Zarehaghi, D.; Maulan, S. Bin Assessing the thermal comfort effects of green spaces: A systematic review of methods, parameters, and plants' attributes. *Sustain. Cities Soc.* **2019**, *49*, 101634. [CrossRef]

99. Hanipah, M.H.; Abdullah, A.H.; Sidik, N.A.C.; Yunus, R.; Yasin, M.N.A.; Yazid, M.N.A.W.M. Assessment of Outdoor Thermal Comfort and Wind Characteristics at Three Different Locations in Peninsular Malaysia. In *Proceedings of the MATEC Web of Conferences*; EDP Sciences: Paris, France, 2016; Volume 47, p. 4005.

100. Höppe, P. Different aspects of assessing indoor and outdoor thermal comfort. *Energy Build.* **2002**, *34*, 661–665. [CrossRef]

101. Huang, T.; Li, J.; Xie, Y.; Niu, J.; Mak, C.M. Simultaneous environmental parameter monitoring and human subject survey regarding outdoor thermal comfort and its modelling. *Build. Environ.* **2017**, *125*, 502–514. [CrossRef]

102. ISO 7730:1994-Moderate Thermal Environments—Determination of the PMV and PPD Indices and Specification of the Conditions for Thermal Comfort. Available online: https://www.iso.org/standard/14567.html (accessed on 29 January 2020).

103. Jain, S.; Garg, V. A review of open loop control strategies for shades, blinds and integrated lighting by use of real-time daylight prediction methods. *Build. Environ.* **2018**, *135*, 352–364. [CrossRef]

104. The Perceived Temperature: The Method of the Deutscher Wetterdienst for the Assessment of Cold Stress and Heat Load for the Human Body. Available online: https://www.semanticscholar.org/paper/ (accessed on 29 January 2020).

105. Jendritzky, G.; De Dear, R.; Havenith, G. UTCI—Why another thermal index? *Int. J. Biometeorol.* **2011**, *56*, 421–428. [CrossRef]

106. Johansson, E. Influence of urban geometry on outdoor thermal comfort in a hot dry climate: A study in Fez, Morocco. *Build. Environ.* **2006**, *41*, 1326–1338. [CrossRef]

107. Johansson, E.; Thorsson, S.; Emmanuel, R.; Krüger, E.L. Instruments and methods in outdoor thermal comfort studies – The need for standardization. *Urban Clim.* **2014**, *10*, 346–366. [CrossRef]

108. Jones, B.W. Capabilities and limitations of thermal models for use in thermal comfort standards. *Energy Build.* **2002**, *34*, 653–659. [CrossRef]

109. Karakounos, I.; Dimoudi, A.; Zoras, S. The influence of bioclimatic urban redevelopment on outdoor thermal comfort. *Energy Build.* **2018**, *158*, 1266–1274. [CrossRef]

110. Lai, D.; Guo, D.; Hou, Y.; Lin, C.; Chen, Q. Studies of outdoor thermal comfort in northern China. *Build. Environ.* **2014**, *77*, 110–118. [CrossRef]

111. Lai, D.; Liu, W.; Gan, T.; Liu, K.; Chen, Q. A review of mitigating strategies to improve the thermal environment and thermal comfort in urban outdoor spaces. *Sci. Total. Environ.* **2019**, *661*, 337–353. [CrossRef]

112. Lai, D.; Zhou, C.; Huang, J.; Jiang, Y.; Long, Z.; Chen, Q. Outdoor space quality: A field study in an urban residential community in central China. *Energy Build.* **2014**, *68*, 713–720. [CrossRef]

113. Li, J.; Niu, J.; Mak, C.M.; Huang, T.; Xie, Y. Assessment of outdoor thermal comfort in Hong Kong based on the individual desirability and acceptability of sun and wind conditions. *Build. Environ.* **2018**, *145*, 50–61. [CrossRef]

114. Li, L.; Zhou, X.; Yang, L. The Analysis of Outdoor Thermal Comfort in Guangzhou during Summer. *Procedia Eng.* **2017**, *205*, 1996–2002. [CrossRef]

115. Lin, P.; Gou, Z.; Lau, S.S.Y.; Qin, H. The Impact of Urban Design Descriptors on Outdoor Thermal Environment: A Literature Review. *Energies* **2017**, *10*, 2151. [CrossRef]

116. Liu, L.; Lin, Y.; Wang, L.; Cao, J.; Wang, D.; Xue, P.; Liu, J. An integrated local climatic evaluation system for green sustainable eco-city construction: A case study in Shenzhen, China. *Build. Environ.* **2017**, *114*, 82–95. [CrossRef]

117. UTCI Calculator. Available online: http://www.utci.org/utcineu/utcineu.php (accessed on 29 January 2020).

118. Mahmoud, A. An analysis of bioclimatic zones and implications for design of outdoor built environments in Egypt. *Build. Environ.* **2011**, *46*, 605–620. [CrossRef]

119. Manavvi, S.; Rajasekar, E. Estimating outdoor mean radiant temperature in a humid subtropical climate. *Build. Environ.* **2020**, *171*, 106658. [CrossRef]

120. Masterton, J.; Richardson, F. *Humidex: A Method of Quantifying Human Discomfort due to Excessive Heat and Humidity*; Environment Canada: Downsview, ON, Canada, 1979.

121. Memon, R.A.; Chirarattananon, S.; Vangtook, P. Thermal comfort assessment and application of radiant cooling: A case study. *Build. Environ.* **2008**, *43*, 1185–1196. [CrossRef]

122. Miao, C.; Yu, S.; Hu, Y.; Zhang, H.; He, X.; Chen, W. Review of methods used to estimate the sky view factor in urban street canyons. *Build. Environ.* **2020**, *168*, 106497. [CrossRef]

123. Middel, A.; Selover, N.; Hagen, B.; Chhetri, N. Impact of shade on outdoor thermal comfort-a seasonal field study in Tempe, Arizona. *Int. J. Biometeorol.* **2016**, *60*, 1849–1861. [CrossRef]

124. Mijorski, S.; Cammelli, S.; Green, J. A hybrid approach for the assessment of outdoor thermal comfort. *J. Build. Eng.* **2019**, *22*, 147–153. [CrossRef]

125. Mittal, H.; Sharma, A.; Gairola, A. A review on the study of urban wind at the pedestrian level around buildings. *J. Build. Eng.* **2018**, *18*, 154–163. [CrossRef]

126. Moran, D.; Health, Y.E.-I. Undefined Evaluation of the environmental stress index (ESI) for hot/dry and hot/wet climates. *Ind. Health* **2006**, *43*, 399–403. [CrossRef]

127. Gonzalez, C.M.; Leon-Rodriguez, A.L.; Navarro-Casas, J. Air conditioning and passive environmental techniques in historic churches in Mediterranean climate. A proposed method to assess damage risk and thermal comfort pre-intervention, simulation-based. *Energy Build.* **2016**, *130*, 567–577. [CrossRef]

128. Abdel-Ghany, A.M.; Al-Helal, I.M.; Shady, M. Human Thermal Comfort and Heat Stress in an Outdoor Urban Arid Environment: A Case Study. *Adv. Meteorol.* **2013**, *2013*, 1–7. [CrossRef]

129. Nagano, K.; Horikoshi, T. New index indicating the universal and separate effects on human comfort under outdoor and non-uniform thermal conditions. *Energy Build.* **2011**, *43*, 1694–1701. [CrossRef]

130. Oh, H.-J.; Jeong, N.-N.; Sohn, J.-R.; Kim, J. Personal exposure to indoor aerosols as actual concern: Perceived indoor and outdoor air quality, and health performances. *Build. Environ.* **2019**, *165*, 106403. [CrossRef]

131. Oliveira, S.; Andrade, H. An initial assessment of the bioclimatic comfort in an outdoor public space in Lisbon. *Int. J. Biometeorol.* **2007**, *52*, 69–84. [CrossRef] [PubMed]

132. Ouali, K.; El Harrouni, K.; Abidi, M.L.; Diab, Y. Analysis of Open Urban Design as a tool for pedestrian thermal comfort enhancement in Moroccan climate. *J. Build. Eng.* **2020**, *28*, 101042. [CrossRef]

133. Pantavou, K.; Santamouris, M.; Asimakopoulos, D.; Theoharatos, G.; Santamouris, M. Empirical calibration of thermal indices in an urban outdoor Mediterranean environment. *Build. Environ.* **2014**, *80*, 283–292. [CrossRef]

134. Pearlmutter, D.; Berliner, P.; Shaviv, E. Integrated modeling of pedestrian energy exchange and thermal comfort in urban street canyons. *Build. Environ.* **2007**, *42*, 2396–2409. [CrossRef]

135. Pearlmutter, D.; Berliner, P.; Shaviv, E. Urban climatology in arid regions: Current research in the Negev desert. *Int. J. Clim.* **2007**, *27*, 1875–1885. [CrossRef]

136. Pickup, J.; Dear, R.D. *An Outdoor Thermal Comfort Index (OUT-SET*)*; 15th ICB ICUC; Macquarie University: Sydney, Australia, 1999.

137. The Heat Index Equation. Available online: https://www.wpc.ncep.noaa.gov/html/heatindex_equation.shtml (accessed on 29 January 2020).

138. Ruiz, M.A.; Correa, E.N. Adaptive model for outdoor thermal comfort assessment in an Oasis city of arid climate. *Build. Environ.* **2015**, *85*, 40–51. [CrossRef]

139. Abu Bakar, A.; Gadi, M.B. Urban Outdoor Thermal Comfort of The Hot-Humid Region. In *Proceedings of the MATEC Web of Conferences*; EDP Sciences: Paris, France, 2016; Volume 66, p. 84.

140. Salata, F.; Golasi, I.; Treiani, N.; Plos, R.; Vollaro, A.D.L. On the outdoor thermal perception and comfort of a Mediterranean subject across other Koppen-Geiger's climate zones. *Environ. Res.* **2018**, *167*, 115–128. [CrossRef]

141. Sharmin, T.; Steemers, K.; Humphreys, M. Outdoor thermal comfort and summer PET range: A field study in tropical city Dhaka. *Energy Build.* **2019**, *198*, 149–159. [CrossRef]

142. Spagnolo, J.; De Dear, R. A field study of thermal comfort in outdoor and semi-outdoor environments in subtropical Sydney Australia. *Build. Environ.* **2003**, *38*, 721–738. [CrossRef]

143. Staiger, H.; Laschewski, G.; Grätz, A. The perceived temperature —A versatile index for the assessment of the human thermal environment. Part A: Scientific basics. *Int. J. Biometeorol.* **2012**, *56*, 165–176. [CrossRef]

144. Sulaiman, H.; Olsina, F. Comfort reliability evaluation of building designs by stochastic hygrothermal simulation. *Renew. Sustain. Energy Rev.* **2014**, *40*, 171–184. [CrossRef]

145. Taleghani, M. Outdoor thermal comfort by different heat mitigation strategies- A review. *Renew. Sustain. Energy Rev.* **2018**, *81*, 2011–2018. [CrossRef]

146. González, A.T.; Chicote, M.A.; García-Ibáñez, P.; Velasco, E.; Rey-Martínez, F.J. Assessing the applicability of passive cooling and heating techniques through climate factors: An overview. *Renew. Sustain. Energy Rev.* **2016**, *65*, 727–742. [CrossRef]

147. Thorsson, S.; Lindqvist, M.; Lindqvist, S. Thermal bioclimatic conditions and patterns of behaviour in an urban park in Göteborg, Sweden. *Int. J. Biometeorol.* **2004**, *48*, 149–156. [CrossRef] [PubMed]

148. Tsitoura, M.; Tsoutsos, T.; Daras, T. Evaluation of comfort conditions in urban open spaces. Application in the island of Crete. *Energy Convers. Manag.* **2014**, *86*, 250–258. [CrossRef]

149. Wan, J.; Yang, K.; Zhang, W.; Zhang, J. A new method of determination of indoor temperature and relative humidity with consideration of human thermal comfort. *Build. Environ.* **2009**, *44*, 411–417. [CrossRef]

150. Ahmed, K.S. Comfort in urban spaces: Defining the boundaries of outdoor thermal comfort for the tropical urban environments. *Energy Build.* **2003**, *35*, 103–110. [CrossRef]

151. Ali, S.B.; Patnaik, S. Thermal comfort in urban open spaces: Objective assessment and subjective perception study in tropical city of Bhopal, India. *Urban Clim.* **2018**, *24*, 954–967. [CrossRef]

152. Aljawabra, F.; Nikolopoulou, M. Influence of hot arid climate on the use of outdoor urban spaces and thermal comfort: Do cultural and social backgrounds matter? *Intell. Build. Int.* **2010**, *2*, 198–217.

153. Amindeldar, S.; Heidari, S.; Khalili, M. The effect of personal and microclimatic variables on outdoor thermal comfort: A field study in Tehran in cold season. *Sustain. Cities Soc.* **2017**, *32*, 153–159. [CrossRef]

154. Baruti, M.; Johansson, E.; Åstrand, J. Review of studies on outdoor thermal comfort in warm humid climates: Challenges of informal urban fabric. *Int. J. Biometeorol.* **2019**, *63*, 1449–1462. [CrossRef]

155. Brager, G.S.; De Dear, R. Thermal adaptation in the built environment: A literature review. *Energy Build.* **1998**, *27*, 83–96. [CrossRef]

156. Brychkov, D.; Garb, Y.; Pearlmutter, D. The influence of climatocultural background on outdoor thermal perception. *Int. J. Biometeorol.* **2018**, *62*, 1873–1886. [CrossRef] [PubMed]

157. Coccolo, S.; Kämpf, J.; Scartezzini, J.-L.; Pearlmutter, D. Outdoor human comfort and thermal stress: A comprehensive review on models and standards. *Urban Clim.* **2016**, *18*, 33–57. [CrossRef]

158. Cohen, P.; Shashua-Bar, L.; Keller, R.; Gil-Ad, R.; Yaakov, Y.; Lukyanov, V.; Bar (Kutiel), P.; Tanny, J.; Cohen, S.; Potchter, O.; et al. Urban outdoor thermal perception in hot arid Beer Sheva, Israel: Methodological and gender aspects. *Build. Environ.* **2019**, *160*, 106169. [CrossRef]

159. Eberhard, J.P. Applying Neuroscience to Architecture. *Neuron* **2009**, *62*, 753–756. [CrossRef]

160. Eberhard, J. *Brain Landscape the Coexistence of Neuroscience and Architecture*; Oxford University Press: New York, NY, USA, 2009.

161. Eliasson, I. The use of climate knowledge in urban planning. *Landsc. Urban Plan.* **2000**, *48*, 31–44. [CrossRef]

162. Fong, C.S.; Aghamohammadi, N.; Ramakreshnan, L.; Sulaiman, N.M.; Mohammadi, P. Holistic recommendations for future outdoor thermal comfort assessment in tropical Southeast Asia: A critical appraisal. *Sustain. Cities Soc.* **2019**, *46*, 101428. [CrossRef]

163. Jabbari, S.G.; Maleki, A.; Kaynezhad, M.A.; Olesen, B.W. Inter-personal factors affecting building occupants' thermal tolerance at cold outdoor condition during an autumn–winter period. *Indoor Built Environ.* **2019**, 1420326X1986799. [CrossRef]

164. Givoni, B.; Noguchi, M.; Saaroni, H.; Pochter, O.; Yaacov, Y.; Feller, N.; Becker, S. Outdoor comfort research issues. *Energy Build.* **2003**, *35*, 77–86. [CrossRef]

165. Groh, J. *Making Space: How the Brain Knows Where Things Are*; Harvard University Press: London, UK, 2014.

166. Haddad, S.; Osmond, P.; King, S. Application of adaptive thermal comfort methods for Iranian schoolchildren. *Build. Res. Inf.* **2016**, *47*, 173–189. [CrossRef]

167. Halawa, E.; Van Hoof, J. The adaptive approach to thermal comfort: A critical overview. *Energy Build.* **2012**, *51*, 101–110. [CrossRef]

168. Hanzl, M.; Ledwoń, S. Analyses of human behaviour in public spaces. Smart Communities. In Proceedings of the ISOCARP/OAPA Congr., Portland, OR, USA, 24–27 October 2017; pp. 653–666.

169. Humphreys, M.; Hancock, M. Do people like to feel 'neutral'? *Energy Build.* **2007**, *39*, 867–874. [CrossRef]

170. Inavonna, I.; Hardiman, G.; Purnomo, A.B. Outdoor thermal comfort and behaviour in urban area. *IOP Conf. Series: Earth Environ. Sci.* **2018**, *106*, 12061. [CrossRef]

171. Jamei, E.; Rajagopalan, P.; Seyedmahmoudian, M.; Jamei, Y. Review on the impact of urban geometry and pedestrian level greening on outdoor thermal comfort. *Renew. Sustain. Energy Rev.* **2016**, *54*, 1002–1017. [CrossRef]

172. Ji, W.; Zhu, Y.; Cao, B. Development of the Predicted Thermal Sensation (PTS) model using the ASHRAE Global Thermal Comfort Database. *Energy Build.* **2020**, *211*, 109780. [CrossRef]

173. Ji, Y.; Wang, Z. Thermal adaptations and logistic regression analysis of thermal comfort in severe cold area based on two case studies. *Energy Build.* **2019**, *205*, 109560. [CrossRef]

174. Ketterer, C.; Matzarakis, A.A. Human-biometeorological assessment of heat stress reduction by replanning measures in Stuttgart, Germany. *Landsc. Urban Plan.* **2014**, *122*, 78–88. [CrossRef]

175. Knez, I.; Thorsson, S. Thermal, emotional and perceptual evaluations of a park: Cross-cultural and environmental attitude comparisons. *Build. Environ.* **2008**, *43*, 1483–1490. [CrossRef]

176. Krüger, E.L.; Kr?ger, E. Impact of site-specific morphology on outdoor thermal perception: A case-study in a subtropical location. *Urban Clim.* **2017**, *21*, 123–135. [CrossRef]

177. Krüger, E.L.; Rossi, F.A. Effect of personal and microclimatic variables on observed thermal sensation from a field study in southern Brazil. *Build. Environ.* **2011**, *46*, 690–697. [CrossRef]

178. Lam, C.K.C.; Lau, K.K.-L. Effect of long-term acclimatization on summer thermal comfort in outdoor spaces: A comparative study between Melbourne and Hong Kong. *Int. J. Biometeorol.* **2018**, *62*, 1311–1324. [CrossRef] [PubMed]

179. Lau, K.K.-L.; Chung, S.C.; Ren, C. Outdoor thermal comfort in different urban settings of sub-tropical high-density cities: An approach of adopting local climate zone (LCZ) classification. *Build. Environ.* **2019**, *154*, 227–238. [CrossRef]

180. Lenzholzer, S.; Klemm, W.; Vasilikou, C. Qualitative methods to explore thermo-spatial perception in outdoor urban spaces. *Urban Clim.* **2018**, *23*, 231–249. [CrossRef]

181. Li, J.; Liu, N. The perception, optimization strategies and prospects of outdoor thermal comfort in China: A review. *Build. Environ.* **2020**, *170*, 106614. [CrossRef]

182. Li, K.; Zhang, Y.; Zhao, L. Outdoor thermal comfort and activities in the urban residential community in a humid subtropical area of China. *Energy Build.* **2016**, *133*, 498–511. [CrossRef]

183. Lin, T.-P. Thermal perception, adaptation and attendance in a public square in hot and humid regions. *Build. Environ.* **2009**, *44*, 2017–2026. [CrossRef]

184. Liu, Z.; Zhao, X.; Jin, Y.; Jin, H.; Xu, X. Prediction of Outdoor Human Thermal Sensation at the Pedestrian Level in High-rise Residential Areas in Severe Cold Regions of China. *Energy Procedia* **2019**, *157*, 51–58. [CrossRef]

185. Malgoyre, A.; Tardo-Dino, P.-E.; Koulmann, N.; Lepetit, B.; Jousseaume, L.; Charlot, K. Uncoupling psychological from physiological markers of heat acclimatization in a military context. *J. Therm. Boil.* **2018**, *77*, 145–156. [CrossRef]

186. Mauree, D.; Naboni, E.; Coccolo, S.; Perera, A.T.D.; Nik, V.M.; Scartezzini, J.-L. A review of assessment methods for the urban environment and its energy sustainability to guarantee climate adaptation of future cities. *Renew. Sustain. Energy Rev.* **2019**, *112*, 733–746. [CrossRef]

187. Metje, N.; Sterling, M.; Baker, C. Pedestrian comfort using clothing values and body temperatures. *J. Wind. Eng. Ind. Aerodyn.* **2008**, *96*, 412–435. [CrossRef]

188. Mi, J.; Hong, B.; Zhang, T.; Huang, B.; Niu, J. Outdoor thermal benchmarks and their application to climate-responsive designs of residential open spaces in a cold region of China. *Build. Environ.* **2020**, *169*, 106592. [CrossRef]

189. Mishra, A.K.; Derks, M.; Kooi, L.; Loomans, M.; Kort, H. Analysing thermal comfort perception of students through the class hour, during heating season, in a university classroom. *Build. Environ.* **2017**, *125*, 464–474. [CrossRef]

190. Mishra, A.K.; Ramgopal, M. Field studies on human thermal comfort—An overview. *Build. Environ.* **2013**, *64*, 94–106. [CrossRef]

191. Nakayoshi, M.; Kanda, M.; Shi, R.; De Dear, R. Outdoor thermal physiology along human pathways: A study using a wearable measurement system. *Int. J. Biometeorol.* **2014**, *59*, 503–515. [CrossRef]

192. Nasir, R.A.; Ahmad, S.; Ahmed, A.Z. Physical Activity and Human Comfort Correlation in an Urban Park in Hot and Humid Conditions. *Procedia Soc. Behav. Sci.* **2013**, *105*, 598–609. [CrossRef]

193. Nikolopoulou, M.; Lykoudis, S. Use of outdoor spaces and microclimate in a Mediterranean urban area. *Build. Environ.* **2007**, *42*, 3691–3707. [CrossRef]

194. Nikolopoulou, M.; Steemers, K. Thermal comfort and psychological adaptation as a guide for designing urban spaces. *Energy Build.* **2003**, *35*, 95–101. [CrossRef]

195. Parkinson, T.; De Dear, R. Thermal pleasure in built environments: Physiology of alliesthesia. *Build. Res. Inf.* **2014**, *43*, 288–301. [CrossRef]

196. Pearlmutter, D.; Jiao, D.; Garb, Y. The relationship between bioclimatic thermal stress and subjective thermal sensation in pedestrian spaces. *Int. J. Biometeorol.* **2014**, *58*, 2111–2127. [CrossRef]

197. Peng, Y.; Feng, T.; Timmermans, H.J.; Nuaa, A. Path analysis of outdoor comfort in urban public spaces. *Build. Environ.* **2019**, *148*, 459–467. [CrossRef]

198. Piselli, C.; Castaldo, V.; Pigliautile, I.; Cabeza, L.F.; Cotana, F. Outdoor comfort conditions in urban areas: On citizens' perspective about microclimate mitigation of urban transit areas. *Sustain. Cities Soc.* **2018**, *39*, 16–36. [CrossRef]

199. Potchter, O.; Cohen, P.; Lin, T.-P.; Matzarakis, A.A. Outdoor human thermal perception in various climates: A comprehensive review of approaches, methods and quantification. *Sci. Total. Environ.* **2018**, 390–406. [CrossRef] [PubMed]

200. Saaroni, H.; Pearlmutter, D.; Hatuka, T. Human-biometeorological conditions and thermal perception in a Mediterranean coastal park. *Int. J. Biometeorol.* **2014**, *59*, 1347–1362. [CrossRef] [PubMed]

201. Salata, F.; Golasi, I.; Proietti, R.; Vollaro, A.D.L. Implications of climate and outdoor thermal comfort on tourism: The case of Italy. *Int. J. Biometeorol.* **2017**, *61*, 2229–2244. [CrossRef]

202. Coburn, A.; Vartanian, O.; Chatterjee, A. Buildings, Beauty, and the Brain: A Neuroscience of Architectural Experience. *J. Cogn. Neurosci.* **2017**, *29*, 1521–1531. [CrossRef]

203. Schweiker, M.; Rissetto, R.; Wagner, A. Thermal expectation: Influencing factors and its effect on thermal perception. *Energy Build.* **2020**, *210*, 109729. [CrossRef]

204. Sharifi, E.; Boland, J. Limits of thermal adaptation in cities: Outdoor heat-activity dynamics in Sydney, Melbourne and Adelaide. *Arch. Sci. Rev.* **2018**, *61*, 191–201. [CrossRef]

205. Sharifi, E.; Boland, J. Passive activity observation (PAO) method to estimate outdoor thermal adaptation in public space: Case studies in Australian cities. *Int. J. Biometeorol.* **2018**, *64*, 231–242. [CrossRef]

206. Shooshtarian, S.; Ridley, I. The effect of physical and psychological environments on the users thermal perceptions of educational urban precincts. *Build. Environ.* **2017**, *115*, 182–198. [CrossRef]

207. Stathopoulos, T.; Wu, H.; Zacharias, J. Outdoor human comfort in an urban climate. *Build. Environ.* **2004**, *39*, 297–305. [CrossRef]

208. Walton, D.; Dravitzki, V.; Donn, M. The relative influence of wind, sunlight and temperature on user comfort in urban outdoor spaces. *Build. Environ.* **2007**, *42*, 3166–3175. [CrossRef]

209. Wilson, E.; Nicol, F.; Nanayakkara, L.; Ueberjahn-Tritta, A. Public Urban Open Space and Human Thermal Comfort: The Implications of Alternative Climate Change and Socio-economic Scenarios. *J. Environ. Policy Plan.* **2008**, *10*, 31–45. [CrossRef]

210. Xi, T.; Wang, Q.; Qin, H.; Jin, H. Influence of outdoor thermal environment on clothing and activity of tourists and local people in a severely cold climate city. *Build. Environ.* **2020**, *173*, 106757. [CrossRef]

211. Xi, T.; Wang, Q.; Wang, S.; Lv, X. Adaptation to outdoor thermal environment of tourists and local people in winter in Harbin. In *Proceedings of the AIP Conference Proceedings*; American Institute of Physics Inc.: College Park, MD, USA, 2019; Volume 2123, p. 020021.

212. Xi, T.; Wang, S.; Wang, Q.; Lv, X. College students' subjective response to outdoor thermal environment in a severely cold climate city. In *Proceedings of the AIP Conference Proceedings*; American Institute of Physics Inc.: College Park, MD, USA, 2019; Volume 2123, p. 020023.

213. Yau, Y.H.; Chew, B.T.; Saifullah, A.Z.A. A Field Study on Thermal Comfort of Occupants and Acceptable Neutral Temperature at the National Museum in Malaysia. *Indoor Built Environ.* **2011**, *22*, 433–444. [CrossRef]

214. Battista, G.; Vollaro, R.D.L.; Zinzi, M. Assessment of urban overheating mitigation strategies in a square in Rome, Italy. *Sol. Energy* **2019**, *180*, 608–621. [CrossRef]

215. Using ENVI-met met BioMet. Available online: http://www.envi-met.info/doku.php. (accessed on 13 April 2020).

216. Imbert, C.; Bhattacharjee, S.; Tencar, J. Simulation of urban microclimate with SOLENE-microclimat —An outdoor comfort case study. *Simul. Ser.* **2018**, *50*, 198–205.

217. Liang, X.; Tian, W.; Li, R.; Niu, Z.; Yang, X.; Meng, X.; Jin, L.; Yan, J. Numerical investigations on outdoor thermal comfort for built environment: Case study of a Northwest campus in China. In *Proceedings of the Energy Procedia*; Elsevier Ltd.: Amsterdam, The Netherlands, 2019; Volume 158, pp. 6557–6563.

218. Lindberg, F.; Holmer, B.; Thorsson, S. SOLWEIG 1.0 – Modelling spatial variations of 3D radiant fluxes and mean radiant temperature in complex urban settings. *Int. J. Biometeorol.* **2008**, *52*, 697–713. [CrossRef]

219. Mackey, C.; Galanos, T.; Norford, L.; Roudsari, M.S.; Bhd, N.S. Wind, Sun, Surface Temperature, and Heat Island: Critical Variables for High-Resolution Outdoor Thermal Comfort Payette Architects. *Proc. 15th IBPSA Conf.* **2017**, 985–993.

220. Matzarakis, A.; Fröhlich, D.; Gangwisch, M.; Ketterer, C.; Peer, A.; Freiburg, A.; Freiburg, D. Developments and applications of thermal indices in urban structures by RayMan and SkyHelios model. In Proceedings of the 9th International Conference on Urban Climate, Toulouse, France, 20–24 July 2015.

221. Matzarakis, A.A.; Rutz, F.; Mayer, H. Modelling radiation fluxes in simple and complex environments: Basics of the RayMan model. *Int. J. Biometeorol.* **2009**, *54*, 131–139. [CrossRef]

222. Matzarakis, A.A.; Rutz, F.; Mayer, H. Modelling radiation fluxes in simple and complex environments—application of the RayMan model. *Int. J. Biometeorol.* **2006**, *51*, 323–334. [CrossRef] [PubMed]

223. Naboni, E.; Coccolo, S.; Meloni, M.; Scartezzini, J.-L. Outdoor Comfort Simulation of Complex Architectural Designs a Review of Simulation Tools from the Designer Perspective. *Build. Perform. Anal. Conf. SimBuild* **2018**, 659–666.

224. Perini, K.; Chokhachian, A.; Dong, S.; Auer, T. Modeling and simulating urban outdoor comfort: Coupling ENVI-Met and TRNSYS by grasshopper. *Energy Build.* **2017**, *152*, 373–384. [CrossRef]

225. Salata, F.; Golasi, I.; Vollaro, R.D.L.; Vollaro, A.D.L. Urban microclimate and outdoor thermal comfort. A proper procedure to fit ENVI-met simulation outputs to experimental data. *Sustain. Cities Soc.* **2016**, *26*, 318–343. [CrossRef]

226. Coccolo, S.; Mauree, D.; Naboni, E.; Kaempf, J.; Scartezzini, J.-L. On the impact of the wind speed on the outdoor human comfort: A sensitivity analysis. *Energy Procedia* **2017**, *122*, 481–486. [CrossRef]

227. Taleb, H.; Taleb, D. Enhancing the thermal comfort on urban level in a desert area: Case study of Dubai, United Arab Emirates. *Urban For. Urban Green.* **2014**, *13*, 253–260. [CrossRef]

228. Tsitoura, M.; Michailidou, M.; Tsoutsos, T. A bioclimatic outdoor design tool in urban open space design. *Energy Build.* **2017**, *153*, 368–381. [CrossRef]

229. Tsoka, S.; Tsikaloudaki, K.; Theodosiou, T. Analyzing the ENVI-met microclimate model's performance and assessing cool materials and urban vegetation applications–A review. *Sustain. Cities Soc.* **2018**, *43*, 55–76. [CrossRef]

230. Fabbri, K.; Di Nunzio, A.; Gaspari, J.; Antonini, E.; Boeri, A. Outdoor Comfort: The ENVI-BUG tool to Evaluate PMV Values Output Comfort Point by Point. *Energy Procedia* **2017**, *111*, 510–519. [CrossRef]

231. Fabbri, K.; Gaspari, J.; Bartoletti, S.; Antonini, E. Effect of facade reflectance on outdoor microclimate: An Italian case study. *Sustain. Cities Soc.* **2020**, *54*, 101984. [CrossRef]

232. Gaspari, J.; Fabbri, K.; Lucchi, M. The use of outdoor microclimate analysis to support decision making process: Case study of Bufalini square in Cesena. *Sustain. Cities Soc.* **2018**, *42*, 206–215. [CrossRef]

233. GhaffarianHoseini, A.; Berardi, U.; GhaffarianHoseini, A.; Al-Obaidi, K. Analyzing the thermal comfort conditions of outdoor spaces in a university campus in Kuala Lumpur, Malaysia. *Sci. Total. Environ.* **2019**, *666*, 1327–1345. [CrossRef] [PubMed]

234. Guan, L.; Bennett, M.; Bell, J. Development of a climate assessment tool for hybrid air conditioner. *Build. Environ.* **2014**, *82*, 371–380. [CrossRef]

235. Huang, K.-T.; Yang, S.-R.; Matzarakis, A.A.; Lin, T.-P. Identifying outdoor thermal risk areas and evaluation of future thermal comfort concerning shading orientation in a traditional settlement. *Sci. Total. Environ.* **2018**, *626*, 567–580. [CrossRef]

236. Huttner, S.; Bruse, M.; Dostal, P. Using ENVI-met to simulate the impact of global warming on the microclimate in central European cities. In Proceedings of the 5th Japanese-German Meeting on Urban Climatology, Freiburg, Germany, 6–8 October 2008.

Review

Energy Poverty and Protection of Vulnerable Consumers. Overview of the EU Funding Programs FP7 and H2020 and Future Trends in Horizon Europe

Danila Longo *, Giulia Olivieri, Rossella Roversi *, Giulia Turci and Beatrice Turillazzi

Architecture Department, University of Bologna, 40136 Bologna, Italy; giulia.olivieri3@unibo.it (G.O.); giulia.turci3@unibo.it (G.T.); beatrice.turillazzi@unibo.it (B.T.)
* Correspondence: danila.longo@unibo.it (D.L.); rossella.roversi@unibo.it (R.R.)

Received: 30 January 2020; Accepted: 21 February 2020; Published: 25 February 2020

Abstract: Energy poverty—involving a combination of factors, such as low household incomes, high energy prices, and low levels of residential energy efficiency—is identified as a complex and increasing issue affecting people's physical health, well-being, and social inclusion. Even though a shared identification of energy poverty is not yet agreed, this phenomenon has been recognized as an EU priority. Several EU legislative documents address the topic, trying to outline its boundaries and provide a framework for mitigative actions. At the same time, different research and demonstration projects have been funded to experiment and evaluate innovative approaches, strategies, and solutions and to promote good practices at national, regional, and local levels. This review paper presents some results of the "ZOOM" project ("Energy zoning for urban systems. Models and relations for the built environment", funded by University of Bologna in the framework of Alma Idea 2017–ongoing), proposing a critical overview of the EU projects directly or indirectly connected to energy poverty—funded under the 7th Framework Program (FP7) and under Horizon 2020 Program (H2020). The aim of such a review is to highlight the main objectives, trends, and related topics of ongoing and concluded projects addressing energy poverty, in order to identify gaps and open issues and to understand the possible orientation and placement of this subject in the future EU research and innovation framework project, Horizon Europe.

Keywords: energy poverty; fuel poverty; vulnerable consumers; vulnerable households; energy vulnerability; energy efficiency; customer engagement; energy citizenship

---

1. Introduction

In recent years, Energy poverty has been recognized as an extensive and increasing issue that is impacting on people's living standards and rights. Consequently, it has attracted growing policy and academic interest in Europe. The European Union embedded such phenomenon in the framework of the "Third Energy Package" [1–3], while the European Economic and Social Committee (EESC) warned about the consequences of the liberalization of energy markets and of the increasing energy vulnerability [4]. Nevertheless, common pan-EU indicators of energy poverty are not yet agreed upon, and various definitions of energy poverty are adopted both in literature and across EU countries [5,6]. Current EU legislation does not require Member States to adopt a common identification of energy poverty, but it does require them to define the concept of 'vulnerable customers'—who may include individuals at risk of or in energy poverty—in order to comply with the requirements stemming from the Third Energy Package.

According to a widely accepted but general description [5,7], energy poverty is a condition when "individuals or households are not able to adequately heat, cool, or provide other required energy services in their homes at affordable cost", where energy services are commonly understood as the benefits of using energy in the home [8]. The EU Energy Poverty Observatory (EPOV) defines energy poverty as the "inability of a household to access socially and materially necessitated levels of energy services in the home" [9]. In this context, "socially necessary" generally means a standard of energy service that allows full participation in society, while the material dimension refers to the health consequences—because inadequately heated or cooled homes have harmful implications for the respiratory, circulatory, and cardiovascular systems, as well as for mental health and well-being [10].

The European Fuel Poverty and Energy Efficiency (EPEE) project, involving five EU Countries (Belgium, France, Italy, Spain, and the United Kingdom), laid the basis for an EU-wide shared definition of energy poverty by proposing to link it to "household's difficulty, sometimes even inability, to adequately heat its dwelling at a fair, income indexed price" [11]. Subsequently, the European Commission defined "energy-poor households" as those with an effective share of energy products expenditure above a predefined threshold, set at "double the national average ratio number" [12]. The United Kingdom is the only country to have defined the threshold: "a household is in a situation of fuel poverty when it has to spend more than 10% of its income on all domestic fuel use, including appliances, to heat the home to a level sufficient for health and comfort" [13]. Given the differences in climate, heating methods, and income assessment, this percentage is not readily applicable to other countries. Therefore, the EPEE consortium proposed that each country adapt this general definition to reflect national characteristics and criteria, while retaining a common view of the problem [11].

This vision is also included in the EU Clean Energy for all Europeans package (CEP), a comprehensive update of the EU energy policy framework to assist the transition towards cleaner energy and to meet the commitments under the Paris Agreement [14]. The CEP rules require that "each EU Country should assess the number of households in energy poverty taking into account the necessary domestic energy services needed to guarantee basic standards of living in the relevant national context" [15] and specific national objectives on energy poverty are to be introduced in the National Energy and Climate Plans (NECPs).

The terms "energy poverty", "fuel poverty", "energy deprivation", "energy vulnerability", "energy precariousness", and "consumer vulnerability" are often used interchangeably in relevant literature and by the EU institutions. The expression 'fuel poverty' has traditionally been used to refer to people in developed countries suffering from inadequate heating in the home; however, in these countries, the importance of other services (cooling, lighting, appliances, IT) has increased substantially in recent years so that heating consumption is no longer a sufficient indicator [16].

According to the latest survey data by the EU Energy Poverty Observatory (EPOV), almost 50 million people in the European Union are affected by energy poverty (source: EU Statistics on Income and Living Conditions, 2016), 44.5 million people were unable to keep their home warm in 2016, and 41.5 million people fell into arrears with their utility bills in 2016 [8]. The condition itself is based on a combination of low household incomes, high energy prices, and low levels of residential energy efficiency. As such, energy poverty does not fully overlap with income poverty, although many low-income households are also energy poor. Often, households suffering from fuel poverty combine low income with an additional degree of vulnerability, such as the elderly, the disabled, and single-parent families.

As pointed out by the European Union Statistics on Income and Living Conditions (EU-SILC) [17], the highest percentage of energy poor households are concentrated in postsocialist Central and Eastern Europe (ECE) countries and in the Mediterranean Sea region, with respectively 20.0% and 16.6% of inadequately heated homes during the winter period against the European average of 12.8% [18].

The number of "vulnerable" households is expected to increase due to rising energy prices [11], climate change, and the transition to green energy. The increase in the average temperature increases the demand for cooling. This trend has already manifested in Mediterranean countries: in Italy, since 2010, the peak demand for electricity in summer is higher than the peak in winter, while in the period 2014–2017, the demand for gas in winter was lower than the average of the previous 15 years [19]. Summertime energy poverty and space cooling difficulties are relatively under-explored: a holistic approach is thus needed. More specifically, it is necessary to combine the assessment of energy poverty considering the whole year and by including all energy services in the home [11].

The energy transition exerts upward pressure on energy prices: the EU electricity prices are affected by the systemic burden of supporting renewable energy and the mechanism for pricing CO_2 emissions; therefore, many researchers predict that EU electricity prices, among the highest in the world, will rise further in the coming decades due to energy transition [20] (European Commission, 2016 [21]). Moreover, the European Economic and Social Committee (EESC) warned of the implications of the liberalization of energy markets and of the current crisis on energy vulnerability [4].

From the late 2000s, several EU legislative documents address the issue of energy poverty, trying to outline its boundaries, provide an overall framework, and lay the ground for actions to mitigate the effects of such phenomenon. Articles 3 of Directive 2009/72/EC (European Parliament and the Council of the European Union, 2009a) and Directive 2009/73/EC (European Parliament and the Council of the European Union, 2009b) require Member States to ensure adequate safeguards to protect vulnerable customers and ask for a definition of the concept of vulnerable customers and appropriate measures to tackle energy vulnerability and energy poverty. EU Directive 2018/2002 (European Parliament and the Council of the European Union, 2018a) requires Member States to account for the need to alleviate energy poverty when designing policy measures to meet their energy saving obligations. Finally, the EU Directive 2019/944 (European Parliament and of the Council of 5 June 2019 on common rules for the internal market for electricity and amending (Directive 2012/27/EU)), referring to EU Regulation 2018/1999 of the European Parliament and of the Council of 11 December 2018, asks Member States to monitor the situation of energy poverty in their countries "taking into account the necessary domestic energy services needed to guarantee basic standards of living in the relevant national context, existing social policy and other relevant policies, as well as Commission indicative guidance on relevant indicators, including geographical dispersion, that are based on a common approach for energy poverty" [22].

The EU has also supported several research and demonstration projects to experiment and evaluate innovative approaches, strategies, and solutions aimed at alleviating energy poverty and promoting good practices at national, regional, and local levels. Nevertheless, the increasing interest in the topic was not matched with a deep investigation of the actual impacts and efficiency of the adopted measures [23]: most of the literature on energy poverty and affordability has focused on their conceptual definition, in particular on the identification of appropriate statistical measures [24].

The review paper presents some results of the ZOOM project ("Energy zoning for urban systems. Models and relations for the built environment"), funded by University of Bologna in the framework of Alma Idea; its objective is to give a critical overview of the EU FP7 and Horizon 2020 funded projects, addressing, directly or indirectly, energy poverty issues, in order to highlight the main goals, trends, related topics, gaps, and open issues. Moreover, the paper aims to understand the possible orientation and placement of the energy poverty subject in the future EU research and innovation framework project, Horizon Europe.

The paper is structured as follows: the "Objective and Methodology" Section 2 outlines the main aims, clarifying that the analysis and selection of energy poverty related projects do not aspire to cover the entirety of scientific production, but are focused on FP7 and H2020 ones. In addition, the section outlines the methodology used to select the projects. In Section 3, dedicated to "Project Results", the systematization and comparison research is illustrated. Section 4, "Open Issues and Future Trends", highlights the relation between FP7 and H2020 and the Horizon Europe framework. In Section 5, conclusions are presented.

2. Objectives and Methodology

The EU funding research and innovation programs are highly committed to addressing energy issues and related topics. This contribution presents a selection of those that are strictly related to energy poverty and the major amount that addresses it collaterally, within more general energy frameworks. The analysis also aims at providing an overview of the possible future placement of the energy poverty subject within the targets and the key strategic orientations of Horizon Europe—the European Union Framework Program for Research and Innovation 2021–2027.

As a matter of fact, projects carried out under the framework programs for research and innovation, including the Intelligent Energy Europe program (IEE) and European Structural and Investment Funds programs [25], have produced knowledge and best practices at various scales that may serve as blueprints for similar initiatives to be implemented in the future, enabling further discussion and uptake in the next research and innovation programs [26].

In this paper, the expression 'energy poverty' has been assumed as preferable: according to Thomson [27], 'energy poverty' is to be preferred to 'fuel poverty', as the first is used in the vast majority of the EU policy documents since 2001.

General reports about the state of play of EU-funded innovation projects on energy poverty in Europe already exist (Joint Research Centre, 2019) [23]; for this reason the present research will focus exclusively on the Horizon 2020 program (H2020, 2014–2020) and the previous Seventh Framework Program (FP7, 2007–2013), because of the specific focus of those projects and, consequently, their high comparability.

During the same period of time, other EU initiatives have been carried out. Those that are mostly focused on energy poverty are:

- Engager—European energy poverty: agenda cocreation and knowledge innovation (2017–2022)—funded by the European Cooperation in Science and Technology organization (COST);
- EPOV—the EU Energy Poverty Observatory (2016–2019)—initiative by the European Commission;
- Evaluate—energy vulnerability and urban transitions in Europe (2013–2018)—funded by the European Research Council;
- EPEE—European fuel poverty and energy efficiency (2006–2009)—IEE funded project.

The list of the projects selected and reported in this paper has been filled in mainly by a keyword search on CORDIS, the Community Research and Development Information Service. CORDIS is the European Commission's primary source for results from the projects funded by the EU's framework programs for research and innovation (FP1 to Horizon 2020). The keywords used in the search were the following (last access 23rd January 2020):

- Energy poverty;
- Fuel poverty;
- Energy vulnerability;
- Vulnerable consumers;
- Vulnerable households.

The keyword search provides hundreds of project results: a multiplicity of project outputs is included as well, such as reports, events, deliverables, and final research summaries. The H2020 and FP7 projects were selected according to their pertinence and the availability of information, resources, documents, and materials, in order to be able to proceed with the knowledge phase and the deepening of their contents.

3. Projects Results—Systematization and Comparison

3.1. Selected FP7 and H2020 Projects Dealing with Energy Poverty Issues

The 43 projects identified throughout ZOOM research address energy-poor issues and implement measures to address causes and solutions (see Appendix A, where all projects are listed and organized in chronological order). At the time of writing, 16 of them are in progress. It must be highlighted that 7 projects, of the total of 16 ongoing projects, are centered on energy poverty, compared to only 5 of the total of 27 completed projects (Figure 1a). This comparison allows us to state that the interest in, and the commitment to, energy poverty from the part of the EU Research and Innovation programs has been considerably increasing since 2017.

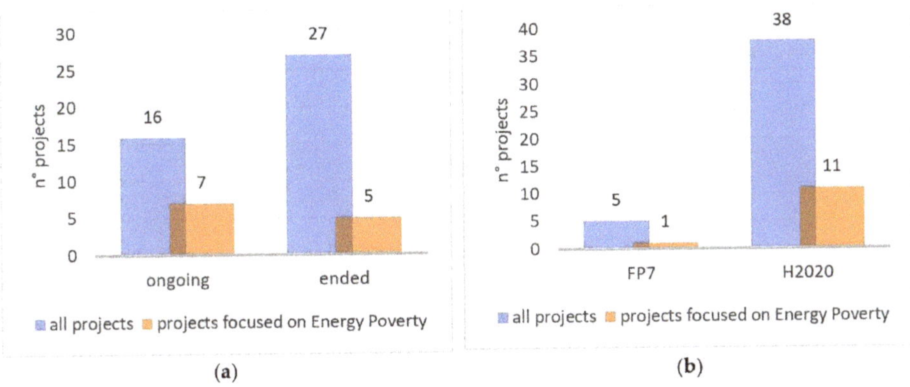

(a)   (b)

Figure 1. (**a**) Number of projects ongoing and ended; (**b**) number of projects per EU Research and Innovation program. Image elaborated by the authors.

This trend is even more evident observing Figure 1b, where the number of projects under FP7 and H2020 are represented: only 1 project has been developed under FP7 out of 11 developed under H2020. Of these, four started in 2019.

As shown in Figure 2, the trend of the number of funded projects over the years is not continuous, but after 2015 it has grown significantly. Especially in 2015, 2016, and 2017, most of the projects were not focused on energy poverty, rather the phenomenon was addressed as a secondary issue or among multiple objectives. In 2019, there is more accentuated attention given to the specific topic.

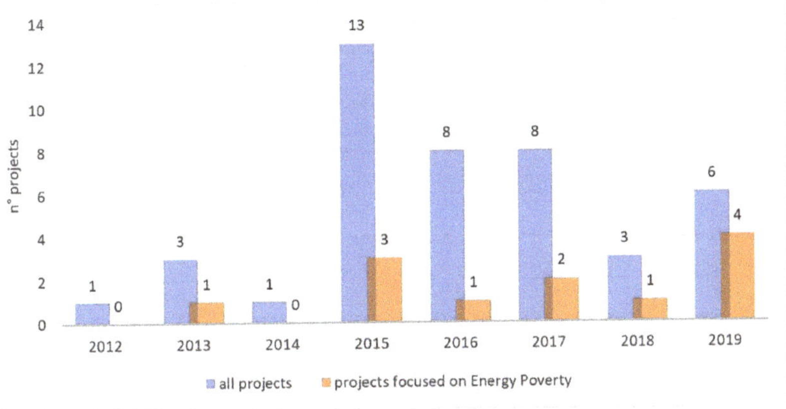

Figure 2. Number of projects per year. Image elaborated by the authors.

3.2. Comparison and Filling Results

After the identification phase, more extensive information on ongoing and completed projects was gathered through dedicated projects' websites, EU online databases, scientific contributions, research outputs and reports, and dissemination and communication materials.

Each selected project has been reported in a factsheet, containing the information set out below in Table 1.

Table 1. Project factsheet content example.

ACRONYM—Description	
Program(s)	Funding scheme
Topic(s)	Start date
Call for proposal	End date
Coordinator	
Partners	
Involved countries	
Target groups	
Objectives	
Methodology	
Expected results	
Outputs application scale	
Energy poverty perspective and/or related topics addressed in the project	
Energy poverty assumed definition	
Indicators	
Case study (if present)	
Geographic position (country/region/city/district)	
Type (private housing/social housing/nonresidential buildings/systems etc.)	
Purpose/Outputs	
Data Sources	Links

As shown in Figure 3, there is a clear prevalence of projects under H2020-EU.3.1.1 "Reducing energy consumption and carbon footprint by smart and sustainable use" and H2020-EU.3.3.7 "Market uptake of energy innovation—building on Intelligent Energy Europe" programs. When interpreting the graph, it should be considered that some projects are financed by two programs and are therefore counted twice. This is the case of three representative projects that appear both in H2020-EU.3.1.1 and in H2020-EU.3.3.7 ("STEP—Solutions to tackle energy poverty", "EmpowerMed—Empowering women to take action against energy poverty in the Mediterranean", and "SocialWatt—Connecting Obligated Parties to Adopt Innovative Schemes towards Energy Poverty"). The three projects are submitted under the call "LC-SC3-EC-2-2018-2019-2020: Mitigating household energy poverty", that is the most specifically addressed at tackling energy poverty. The proposed actions are required to cover at least one of the following objectives [28]:

- facilitating behavioral change and implementation of low-cost energy efficiency measures tailored for energy poor households;
- supporting the set-up of financial and nonfinancial support schemes for energy efficiency and/or small-scale renewable energy investments for energy poor households;
- developing, testing, and disseminating innovative schemes for energy efficiency/renewable electricity sources (RES) investments established by utilities or other obligated parties under Article 7 (of Energy Efficiency Directive 2012/27/EU) [29]. New projects are expected in the coming years, funded by the Calls for Proposals 2019 and 2020.

As shown in Figure 4, the most represented funding scheme is the Coordination and Support Actions (CSA), dealing with the coordination and networking of research and innovation projects, programs, and policies whose funds for research and innovation per se are covered by other funding sources.

Only one project centered on energy poverty is a Research and Innovation Actions (RIA) project, leading to the development of new knowledge or a new technology. None are Innovation Action (IA) projects, with a closer focus on market activities, aimed at producing or improving products or services. Considering the projects that tackle energy poverty as an implication of the measures designed for other issues, the number of RIA is relevant.

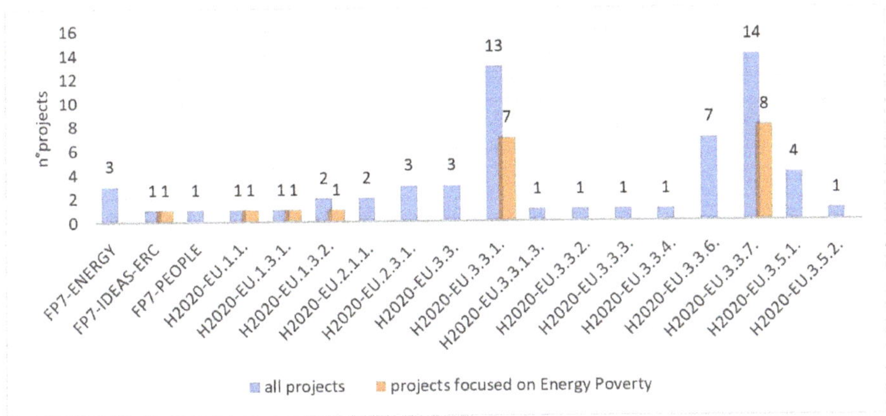

Figure 3. Number of projects per funding program. Image elaborated by the authors.

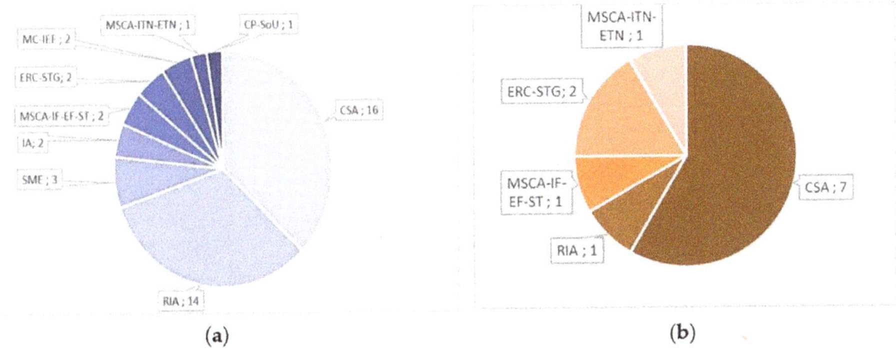

(a)   (b)

Figure 4. (a) Number of projects per funding scheme; (b) number of projects focused on energy poverty per funding scheme (CSA—coordination and support action; RIA—Research and Innovation Action; SME—Small and Medium-sized Enterprises instrument phase; IA—innovation action; MSCA-IF-EF-ST—Marie Sklodowska Curie Actions—Individual Fellowships—Standard European Fellowships; ERC-SG—European Research Council—Starting Grant; MC-IEF—Intra-European Fellowships (IEF); CP-SoU—Collaborative Project—Scale of Unit; MSCA-ITN-ETN—Innovative Training Networks—European Training Networks. Image elaborated by the authors.

The overall budget of allocated funds is 121,275,563 euro and the prevalent amount range is between 500,000 and 2,000,000 euro.

The geographic distribution of project leaders in EU countries, as well as the distribution of participants stakeholders, is mainly concentrated in United Kingdom, Italy, Germany, and Spain (see Figure 5a). The United Kingdom confirms its centrality in the energy poverty debate and has developed a growing number of national actions and policy frameworks. Although energy poverty affects more consistently the post-socialist Central and Eastern Europe (ECE) countries, only two related projects were developed by countries from this area: EnergyKeeper, coordinated in Lithuania and dealing with the development of an intelligent storage solution to optimize energy demand and supply of isolated communities, and SOLACE 'System of production and delivery self-assembly, carbon neutral, energy positive, and income generating house, which reduces housing overburden declining inequality gap', coordinated in Polonia and facing the challenge of rethinking house construction to allow people in need to live in a decent and sustainable way.

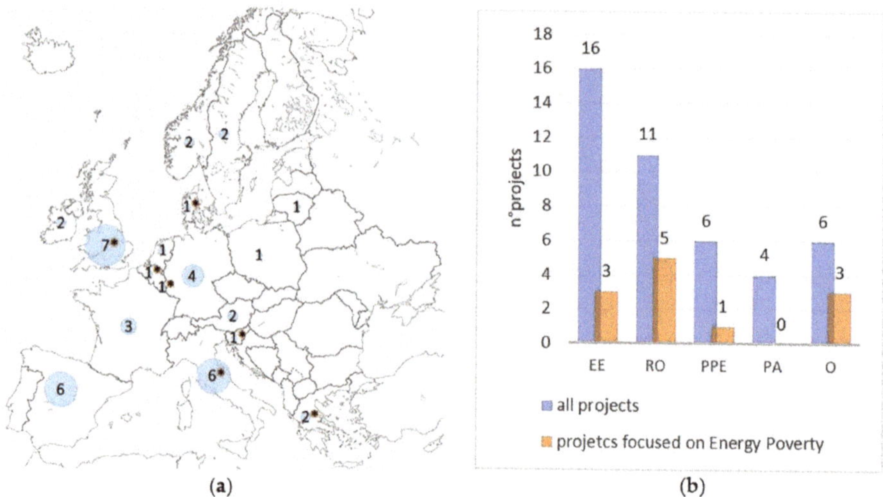

Figure 5. (**a**) Project leader distribution (* pointed projects specifically focused on energy poverty); (**b**) number of projects per type of coordinator (EE—education establishments; RO—research organizations; PPE—private for-profit entities; PA—public administrations; and O—others). Image elaborated by the authors.

The coordination of projects is largely entrusted to universities and research organizations (see Figure 5b).

As shown in Figure 6, most of the projects are addressed at consumers, but only a few are specifically tailored to set practical energy efficiency measures helping households at risk or in energy poverty conditions. A significant percentage of projects also involve public administrations and policy makers, while a minor percentage comprises energy utilities whose presence is more frequent as project partners.

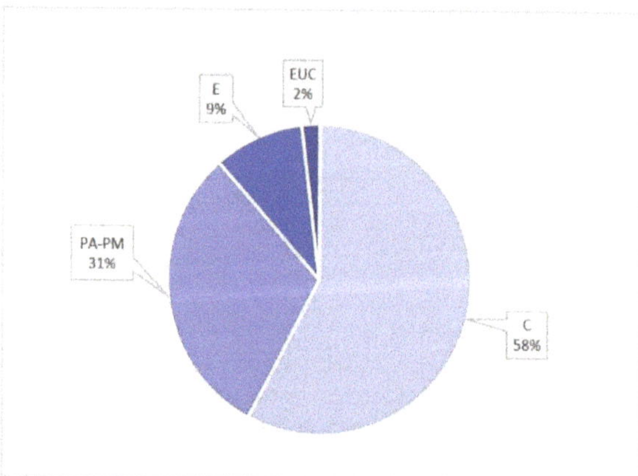

Figure 6. Percentage of projects per addressed target group. (C—consumers; PA-PM—public administrations and policy makers; E—entrepreneurs; and EUC—energy utility companies). Image elaborated by the authors.

The objectives of the surveyed projects are often multiple and combined. In general, the projects show a lack of specific attention to the needs of vulnerable consumers and the wider social aspects of energy poverty (see Figure 7a).

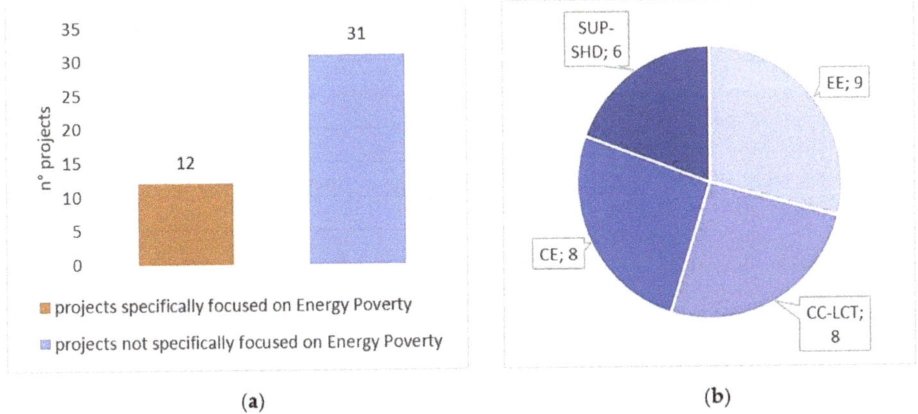

Figure 7. (**a**) Number of projects specifically focused on energy poverty and projects not specifically focused on it; (**b**) main issues addressed by projects not specifically focused on energy poverty (CC-CET—climate change and low carbon transition; EE—energy efficiency; CE—citizens engagement; SUP-SHD—sustainable urban planning and social housing districts). Image elaborated by the authors.

The significant number of detected pertinent projects has allowed us to split energy poverty-centered projects from nonenergy poverty centered ones. For the latter category, the main topics that energy poverty has been coupled to or that feature energy poverty as a related impact were identified and grouped as follows:

- Citizens engagement—involvement of vulnerable consumers in order to improve their awareness, to guide them in the adoption of behavioral changes aimed at reducing energy consumption and to support them in the digitization process;
- Climate change and low-carbon transition—policies and actions to tackle climate change and its impacts without negative repercussions on vulnerable citizens; initiatives to fulfil the key objectives of the European Strategic Energy Technology Plan (SET Plan);
- Energy efficiency—developing and monitoring technological solutions for Near Zero Energy Buildings (NZEB) and building retrofitting. Solutions involve the development of plant systems, building components, and technological systems to reduce consumption and generate energy savings;
- Sustainable urban planning and social housing districts—strategies towards urban planning renovation, regeneration of public spaces, social housing and vulnerable districts, and the pursuit of nearly zero energy cities.

In Figure 7b, the 31 non-focused projects are divided according to the groups of related abovementioned topics: those dealing with citizen engagement, climate change, and low-carbon transition are approximately in equivalent number.

As energy poverty is mainly the product of the energy efficiency of the house, combined with the cost of heating fuel and the household income, Member States have promoted a range of policy measures to move the majority of the population out of fuel poverty involving these three factors and generating the most relevant instruments and practices assessed in the literature [10]. A first category of measures includes financial support and social tariffs, targeted for groups of vulnerable consumers, such as low-income citizens, single parents, over-consumer households, large families, the unemployed, or the retired. Reducing energy poverty through financial support does not tackle CO2 emissions as it allows people living in poor quality houses to use more energy to reach a comfortable thermal level, thereby increasing carbon emissions. A second category includes, on the one hand, household energy efficiency solutions to improve envelopes, heating/cooling systems, and household appliances, and on the other, the possibility of obtaining subsidies, grants, or tax reductions for energy efficiency improvements and investments. Making homes more energy efficient is a long-term sustainable solution, which allows people to use less energy to heat/cool their homes adequately with a positive impact on carbon emissions. For this reason, energy efficiency is recognized as an important driver in tackling energy poverty [30].

The H2020 projects dealing with climate change and low-carbon transition often intersect those dealing with sustainable urban planning and social housing districts. The common ground is the Strategic Energy Technology Plan (SET-Plan) that in key-action three focuses on new technologies and services for consumers and is composed by SET-Plan 3.1 "smart solutions for energy consumers" and SET-Plan 3.2 "smart cities and communities". Action 3.1 aims at the creation of a digitalized eco-system where consumers, companies, or stakeholders can offer or use energy services for houses and commercial buildings in cities through different sources of data, using real-time monitoring and control [31]; Action 3.2 aspires to create the conditions for planning, deployment, and replication of 100 Positive Energy District (PED) by 2025 for sustainable urbanization [32].

4. Open Issues and Future Trends

The main open issue and object of debate is still the definition of energy poverty itself, its conceptual identification, and the detection of appropriate statistical measures (Hills, 2012) [24]. Academics have different opinions about the need of a common definition: Thomson et al. [33] and Dobbins et al. [34], argue that a common definition and approach could facilitate the recognition of the problem at EU level, paving the way for more detailed national definitions and encouraging synergies between Member States. Others [35], among them Grevisse et al. [36], argue that, given the significant context differences, a common EU definition is undesirable, and the choice of energy poverty definitions and policies should rest with Member States [11]. In particular, the identification of an absolute threshold for energy poverty at the European level has turned out to be very challenging, making it preferable to develop more regionally specific and targeted settlement-level data [8] (p.34). For this reason, the EU Directive 2019/944 and the EU Regulation 2018/1999 state that energy poverty must be evaluated taking into account the necessary domestic energy services needed to guarantee basic standards of living in the relevant national context [22].

In the examined projects, the limited availability of relevant and comparable data, often collected for other purposes, makes it difficult to get quantitative and measurable results. Moreover, projects often use different methodologies to estimate or measure results; therefore, shared and standardized monitoring, evaluation, and reporting is needed in order to identify and provide evidence of successful solutions and to boost their scaling up and replication [23] (p.53). The STEP-IN project suggests the adoption of the EU Energy Poverty Observatory (EPOV) indicators, divided into four primary indicators (arrears on utility bills, low absolute energy expenditure, high share of energy expenditure in income, and inability to keep home adequately warm) and secondary indicators, related to energy poverty but not directly addressing energy poverty itself, e.g., energy prices and dwellings-related data. According to the ASSIST'Support Network for Household Energy Saving Smart Ventilation Control' project, the more widespread indicator is the Low Income High Costs indicator (LIHC): low income high costs, defined by Hills in 2012 [24]. For Belgium, Poland, Spain, Finland, and the UK this indicator was already available. In Italy, there are data and statistics about income and energy expenses; however, currently they cannot be correlated, thus, depending on the different goals, different indicators are in use. Among them, an adapted LIHC indicator has been proposed [37].

Another open issue is the development and implementation of a consumer engagement strategy: in many selected projects, the engagement of energy-poor/vulnerable households is considered of outmost importance to the achievement of the project's objectives. The social, cultural, and behavioral perspective of energy poverty is becoming more and more central. Citizen empowerment and engagement is at the core of the 'Clean Energy for all Europeans (CEP)' whose purpose is to ensure a clean and fair energy transition at all levels of the economy. The CEP attributes a crucial role to citizens and community activities, introducing the energy community into the regulatory framework. According to the CEP, the democratization of energy will alleviate energy poverty and protect vulnerable citizens [14] (p.12). Strategies aimed at supporting consumer awareness, participation, and trust and building confidence and sense of community emerge as a mainstream and cross-cutting issue, to be integrated into projects, actions, and policy. An ongoing representative pilot project is the GECO—Green Energy Community, financed by Climate-KIC, coordinated by the Modena Energy and Sustainable Development Agency (AESS), ENEA, and the University of Bologna [38]. GECO started in July 2019 and aims to promote the production and self-consumption of renewable energy in two districts of Bologna through the creation of a local energy community, involving local inhabitants, commercial activities, and companies [39].

As the H2020 program will expire in the present year, the new European Union Framework Program for Research and Innovation 2021–2027 is being completed (Horizon Europe). The mapping of relevant H2020 calls and related projects has been a step of the development construction process of the strategic plan for Horizon Europe, helping addressing key challenges. Among them, behavioral factors and participative practices are confirmed crucial issues, in particular those using co-designing solutions, building on vulnerable consumers' real needs and expectations.

In the currently available Horizon Europe documents, in particular those discussed within the ECTP innovative built environment platform—EeB committee, energy poverty is not explicitly mentioned, but some of the main open issues are included in the draft clusters. According to the 31 October 2019 draft of the orientation document (orientations towards the first strategic plan for Horizon Europe revised following the co-design process) [40], cluster 1, 'health', is devoted to identifying the environmental and social health determinants in order to improve the understanding of health drivers and risk factors generated by the social, economic, and physical environment. It aims also to promote healthy lifestyles and consumption behavior, with special attention to vulnerable and disadvantaged people. Cluster 2, 'culture, creativity, and inclusive society', will support actions to tackle social, economic, and political inequalities; cluster 4, "digital, industry, and space", will contribute to developing a low-carbon, circular, and clean industry and fostering societal engagement in the use of technologies; and cluster 5, 'climate, energy, and mobility', will contribute to the technological, economic, and societal transformations required to achieve climate neutrality, ensuring at the same time a socially fair transition.

5. Conclusions

In spite of the European legislation efforts to explicitly recognize the energy poverty issue and provide a common framework for the protection and improvement of the conditions of vulnerable consumers, these efforts have not yet been converted into agreed measures, lacking common EU-wide understanding of the phenomenon, while policies to be adopted to pursue this objective are far from plain and unambiguous.

The heterogeneity of interventions reflects the terminological uncertainty, motivated also by the lack of common parameters to define 'energy poverty' at EU level and the variety of causes of energy poverty. These causes can be very different, depending on the various contexts of the considered countries. Consequently, it is quite difficult to define common indicators and to find relevant quantitative data to characterize on a same basis the different national situations. Moreover, as it has already been pointed out, many projects pursue multiple objectives that complement each other but often compete for priorities and budget.

The JRC report stated that "time-trend data for all project categories suggest that the growing attention attracted by energy poverty at policy level in recent years has not yet been reflected in the research and innovation initiatives carried out to date with EU financial support" [23] (p.12), proving how more projects with a clearer focus on energy poverty and vulnerable consumers would improve the understanding of this phenomenon and help identify effective solutions to address it. The recognition of EU Research and Innovation projects reported in the present paper, only including H2020 and FP7 funded projects, shows that 2019 has been a turning point year: after the JRC publication, three projects have been funded under the H2020 topic LC-SC3-EC-2-2018-2019-2020—mitigating household energy poverty" and new projects are further expected in the present year as the call for proposal is still open.

FP7 and H2020 programs leave a considerable number of open issues about energy poverty and protection of vulnerable consumers. The geographic prevalence of project case studies in the north west countries and the insufficient theoretical and applicative studies in the north east and Mediterranean countries, where energy poverty affects a considerable percentage of the population, still remain a weakness. Only a few funded projects address the issue of energy poverty from a geographical point of view, focusing on the most vulnerable areas. One is EVALUATE 'Energy Vulnerability and Urban Transitions in Europe', funded in 2013 under FP7 program, operating in the post-socialist countries of Central and Eastern Europe. A more recent one is EmpowerMed 'Empowering women to tackle action against energy poverty in the Mediterranean Sea region', funded in 2019 and focusing on the engagement and involvement of women as main operators in the management of households. Both the behavioral and the gender perspective are worthy of further investigation and Horizon Europe is expected to head in this direction.

Another interesting and promising result emerging from the project review is the cross cutting and multidisciplinary approach: many projects address energy poverty in an integrated way by merging the spatial dimension with social and technological experimentations. However, a significant number of projects address only technological or energy issues through the development of specific building components, thus confining their contribution to the building scale.

In conclusion, the review of FP7 and H2020 funded projects demonstrates that the EU Research and Innovation initiatives are developing a better understanding of the types and needs of energy-poor households and are demonstrating innovative strategies to address energy poverty. Their piloting innovative solutions on vulnerable consumers, despite the aforementioned limits and shortcomings, help to face and anticipate problems, in order to catch the opportunities and build/create inclusive energy communities.

Author Contributions: Conceptualization, D.L., R.R. and B.T.; methodology, R.R.; software, G.T.; validation, R.R. and B.T.; formal analysis, G.O.; investigation, G.O. and G.T.; resources, D.L., B.T. and R.R.; data curation, G.O. and G.T.; writing—original draft preparation, R.R.; writing—review and editing, D.L. and R.R.; visualization, G.O. and G.T.; supervision, D.L.; project administration, D.L. and B.T.; and funding acquisition, D.L. All authors have read and agreed to the published version of the manuscript.

Funding: This research received no external funding.

Conflicts of Interest: The authors declare no conflict of interest.

Appendix A

Table A1. List of FP7 and H2020 selected projects.

n°	Acronym	Period	Description
01	STEP-UP	2012–2015	Strategies Towards Energy Performance and Urban Planning
02	EVALUATE *	2013–2018	Energy Vulnerability and Urban Transitions in Europe
03	R2CITIES	2013–2017	Renovation of Residential urban spaces: towards nearly zero energy cities
04	PACE	2013–2015	Property Assessed Clean Energy
05	INSIGHT_E	2014–2017	Interdisciplinary Strategic Intelligence wareHouse and Think-tank for Energy
06	REMOURBAN	2015–2019	REgeneration MOdel for accelerating the smart URBAN transformation
07	USES *	2015–2018	Understanding Social-Ecological Systems: Coupling population and satellite remotely sensed environmental data to improve the evidence base for sustainable development
08	SMARTESS	2015–2021	Social innovation Modelling Approaches to Realizing Transition to Energy Efficiency and Sustainability Innovate
09	GREEN-WIN	2015–2018	Green growth and win-win strategies for sustainable climate action
10	QBOT	2015–2017	Fast and economic insulation of buildings using robotic systems
11	CD-LINKS	2015–2019	Linking Climate and Development Policies—Leveraging International Networks and Knowledge Sharing
12	ENTRUST	2015–2018	Energy System Transition Through Stakeholder Activation, Education, and Skills Development
13	SMART-UP	2015–2018	Vulnerable consumer empowerment in a smart meter world
14	EnerGAware	2015–2018	Energy Game for Awareness of energy efficiency in social housing communities
15	COMBI	2015–2018	Calculating and Operationalizing the Multiple Benefits of Energy Efficiency Improvements in Europe
16	Decent Living Energy *	2015–2019	Energy and emissions thresholds for providing decent living standards to all
17	LEMON *	2015–2020	LEMON Less Energy More OpportuNities
18	TRIBE	2015–2018	TRaIningBehaviors towards Energy efficiency: Play it!
19	REEM	2016–2019	Role of technologies in an energy efficient economy—model-based analysis of policy measures and transformation pathways to a sustainable energy system
20	Think Nature	2016–2019	Renovation of Residential urban spaces: Towards nearly zero energy CITIES
21	ABRACADABRA	2016–2019	Assistant Buildings' addition to Retrofit, Adopt, Cure, and Develop the Actual Buildings up to zeRo energy, Activating a market for deep renovation
22	ECHOES	2016–2019	Energy CHOices supporting the Energy union and the Set-plan
23	EnerSHIFT *	2016–2020	Energy Social Housing Innovative Financing Tender
24	ENERGISE	2016–2019	European Network for Research, Good Practice, and Innovation for Sustainable Energy
25	ENABLE.EU	2016–2019	Enabling the Energy Union through understanding the drivers of individual and collective energy choices in Europe
26	INNOPATHS	2016–2020	Innovation pathways, strategies, and policies for the Low-Carbon Transition in Europe
27	SAVES2 *	2017–2020	Students Achieving Valuable Energy Savings 2

Table A1. *Cont.*

n°	Acronym	Period	Description
28	EnergyKeeper	2017– 2019	Keep the Energy at the right place!
29	ASSIST *	2017–2020	ASSIST - Support Network for Household Energy Saving
30	AirEX	2017–2018	Smart Ventilation Control
31	SOLACE	2017–2017	System of production and delivery self-assembly, carbon neutral, energy positive, and income generating house, which reduces housing overburden declining inequality gap
32	AZEB	2017–2020	Affordable zero energy buildings
33	InBETWEEN	2017–2020	ICT enabled BEhavioral change ToWards Energy EfficieNt lifestyles
34	PV-Prosumers4Grid	2017–2020	Development of innovative self-consumption and aggregation concepts for PV Prosumers to improve grid load and increase market value of PV
35	TRANSFAIR	2018–2020	Unfair transitions? A critical examination low-carbon energy pathways in the EU from a domestic energy vulnerability perspective
36	STEP-IN *	2018–2020	Using Living Labs to roll out Sustainable Strategies for Energy Poor Individuals
37	SCORE	2018–2021	Supporting Consumer Co-Ownership in Renewable Energies
38	STEP *	2019–2021	Solution to Tackle Energy Poverty
39	EmpowerMed *	2019–2023	Empowering women to take action against energy poverty in the Mediterranean
40	SocialWatt *	2019–2022	Connecting Obligated Parties to Adopt Innovative Schemes towards Energy Poverty
41	ODYSSEE-MURE	2019–2021	Monitoring EU energy efficiency first principle and policy implementation
42	Smart BEEjs *	2019–2023	Human-Centric Energy Districts: Smart Value Generation by Building Efficiency and Energy Justice for Sustainable Living
43	HIROSS4all	2019–2022	Home integrated renovation one-stop-shop for vulnerable districts

* Projects specifically focused on energy poverty.

References

1. European Parliament. *Directive 2009/72/EC of the European Parliament and of the Council of 13 July 2009 Concerning Common Rules for the Internal Market in Electricity and Repealing Directive 2003/54/EC*; European Parliament: Brussels, Belgium, 2009.
2. European Parliament. *Directive 2009/73/EC of the European Parliament and of the Council of 13 July 2009 Concerning Common Rules for the Internal Market in Natural Gas and Repealing Directive 2003/55/EC*; European Parliament: Brussels, Belgium, 2009.
3. Third Energy Package. 2009. Available online: https://ec.europa.eu/energy/en/topics/markets-and-consumers/market-legislation (accessed on 29 December 2019).
4. European Economic and Social Committee (EESC). Opinion of the European Economic and Social Committee 'for Coordinated European Measures to Prevent and Combat Energy Poverty'. *Off. J. Eur. Union* **2013**, *56*, 1–8. [CrossRef]
5. Trinomics, Selecting Indicators to Measure Energy Poverty. 2016. Available online: https://ec.europa.eu/energy/sites/ener/files/documents/Selecting%20Indicators%20to%20Measure%20Energy%20Poverty.pdf (accessed on 28 December 2019).
6. Insight_E. Policy Report, Energy Poverty and Vulnerable Consumers in the Energy Sector across the EU: Analysis of Policies and Measures. 2015. Available online: https://ec.europa.eu/energy/sites/ener/files/documents/INSIGHT_E_Energy%20Poverty%20-%20Main%20Report_FINAL.pdf (accessed on 18 January 2020).
7. Bouzarovski, S. *Energy Poverty. (Dis)Assembling Europe's Infrastructural Divide*, 1st ed.; Palgrave Macmillian: London, UK, 2018. [CrossRef]
8. Thomson, H.; Bouzarovski, S. *Addressing Energy Poverty in the European Union: European State of Play and Action*; EU Energy Poverty Observatory (EPOV): Brussels, Belgium, 2019; Available online: https://www.energypoverty.eu/sites/default/files/downloads/observatory-documents/19-06/paneureport2018_updated2019.pdf (accessed on 20 January 2020).
9. Bouzarovski, S. Energy Poverty in the European Union: Landscapes of Vulnerability. *WIREs Energy Environ.* **2014**, *3*, 276–289. [CrossRef]
10. Fowler, T.; Southgate, R.J.; Waite, T.; Harrell, R.; Kovats, S.; Bone, A.; Murray, V. The Health Impacts of Cold Homes and Fuel Poverty. *Eur. J. Public Health* **2015**, *25*, 339–345. [CrossRef] [PubMed]
11. European Fuel Poverty and Energy Efficiency Project (EPEE). 2009. Available online: https://ec.europa.eu/energy/intelligent/projects/sites/iee-projects/files/projects/documents/epee_european_fuel_poverty_and_energy_efficiency_en.pdf (accessed on 18 January 2020).
12. European Commission. An Energy Policy for Customers. In *Commission Staff Working Paper*; European Commission: Brussels, Belgium, 2010. Available online: https://ec.europa.eu/energy/sites/ener/files/documents/sec%282010%291407.pdf (accessed on 15 January 2020).
13. Boardman, B. *Fuel Poverty: From Cold Homes to Affordable Warmth*; Belhaven Press: London, UK, 1991.
14. European Commission. *Clean Energy for All European*; Publications Office of the European Union: Brussels, Belgium, 2019; Available online: https://op.europa.eu/en/publication-detail/-/publication/b4e46873-7528-11e9-9f05-01aa75ed71a1/language-en?WT.mc_id=Searchresult&WT.ria_c=null&WT.ria_f=3608&WT.ria_ev=search (accessed on 27 January 2020).
15. European Commission. United in Delivering the Energy Union and Climate Action-Setting the Foundations Fora Successful Clean Energy Transition. 2019. Available online: https://eur-lex.europa.eu/legal-content/EN/TXT/PDF/?uri=CELEX:52019DC0285&from=EN (accessed on 22 January 2020).
16. Bouzarovski, S.; Petrova, S. A Global Perspective on Domestic Energy Deprivation: Overcoming the Energy Poverty-Fuel Poverty Binary. *Energy Res. Soc. Sci.* **2015**, *10*, 31–40. [CrossRef]
17. European Union Statistics on Income and Living Conditions (EU-SILC). Available online: https://ec.europa.eu/eurostat/web/microdata/european-union-statistics-on-income-and-living-conditions (accessed on 17 February 2020).
18. Simcock, N.; Petrova, S. Energy poverty and vulnerability: A geographic perspective. In *Handbook on the Geographies of Energy*; Solomon, B.D., Calvert, K.E., Eds.; Edward Elgar: Cheltenham, UK, 2017; pp. 425–437.

19. Faiella, I.; Lavecchia, L.; Miniaci, R.; Valbonesi, P. Rapporto OIPE sulla povertà energetica in Italia. In Proceedings of the La Povertà Energetica in Italia: Come Misurarla e Come Combatterla, Milano, Italy, 4 June 2019.

20. IEA. World Energy Outlook. 2017. Available online: https://www.iea.org/reports/world-energy-outlook-2017 (accessed on 10 January 2020).

21. European Commission. EU Reference Scenario 2016: Energy Transport and GHG Emissions Trends to 2050. Available online: https://ec.europa.eu/energy/sites/ener/files/documents/20160713%20draft_publication_REF2016_v13.pdf (accessed on 22 January 2020).

22. EU Regulation 2018/1999 of the European Parliament and of the Council of 11 December 2018, on the Governance of the Energy Union and Climate Action, Official Journal of the European Union 328/1, point (d) of Article 3(3) of Regulation (EU) 2018/1999. Available online: https://eur-lex.europa.eu/legal-content/EN/TXT/PDF/?uri=CELEX:32018R1999&from=EN (accessed on 21 January 2020).

23. Gangale, F.; Mengolini, A. *Energy Poverty through the Lens of EU Research and Innovation Projects*; Publications Office of the European Union: Luxembourg, Belgium, 2019; p. 2. [CrossRef]

24. Hills, J. *Getting the Measure of Fuel Poverty: Final Report of the Fuel Poverty Review*; CASEreport, 72; London School of Economics and Political Science: London, UK, 2012. Available online: https://eprints.lse.ac.uk/43153/1/CASEreport72%28lsero%29.pdf (accessed on 10 January 2020).

25. European Structural and Investment Funds (ESIF). Available online: https://ec.europa.eu/info/funding-tenders/funding-opportunities/funding-programmes/overview-funding-programmes/european-structural-and-investment-funds_en (accessed on 22 January 2020).

26. Report on the First Results of H2020 Projects on Energy Efficiency and System Integration-Final Report. 2017. Available online: https://ec.europa.eu/energy/sites/ener/files/documents/ed62228_h2020_energy_evaluation_final_report_v1.5_3_0.pdf (accessed on 22 January 2020).

27. EU Energy Poverty Observatory (EPOV). Brief Overview of the EU Discourse on Fuel Poverty and Energy Poverty. 2014. Available online: https://www.energypoverty.eu/news/brief-overview-eu-discourse-fuel-poverty-energy-poverty (accessed on 18 January 2020).

28. Horizon 2020 Work Programme 2018–2020, Secure, Clean and Efficient Energy. Available online: https://ec.europa.eu/research/participants/data/ref/h2020/wp/2018-2020/main/h2020-wp1820-energy_en.pdf (accessed on 20 January 2020).

29. European Parliament. *Directive 2012/27/EU of the European Parliament and of the Council of 25 October 2012 on Energy Efficiency, Amending Directives 2009/125/EC and 2010/30/EU and Repealing Directives 2004/8/EC and 2006/32/EC*; European Parliament: Brussels, Belgium, 2012.

30. Guidance Document on Vulnerable Consumers. 2013. Available online: https://ec.europa.eu/energy/sites/ener/files/documents/20140106_vulnerable_consumer_report_0.pdf (accessed on 20 January 2020).

31. Temporary Working Group on Smart Solutions for Energy Consumers, SET-Plan 3.1 Energy Consumers Implementation Plan. 2018. Available online: https://setis.ec.europa.eu/system/files/set_plan_consumers_implementation_plan.pdf (accessed on 15 January 2020).

32. Temporary Working Group on Smart Cities and Communities, SET-Plan 3.2 Europe to Become a Global Role Model in Integrated, Innovative Solutions for the Planning, Deployment, and Replication of Positive Energy Districts. 2018. Available online: https://setis.ec.europa.eu/system/files/setplan_smartcities_implementationplan.pdf (accessed on 20 January 2020).

33. Thomson, H.; Snell, C.; Liddell, C. Fuel poverty in the European Union: A concept in need of definition. *People Place Policy* **2016**, *10*, 5–24. [CrossRef]

34. Dobbins, A.; Fuso Nerini, F.; Deane, P.; Pye, S. Strengthening the EU response to energy poverty. *Nat. Energy* **2019**, *4*, 2–5. [CrossRef]

35. Deller, D. Energy affordability in the EU: The risks of metric driven policies. *Energy Policy* **2018**, *119*, 168–182. [CrossRef]

36. Grevisse, F.; Brynart, M. Energy poverty in Europe: Towards a more global understanding. *Eur. Counc. Energy Effic. Econ.* **2011**, *71*, 537–549.

37. Faiella, L.; Lavecchia, L.; Borgarello, M. Una nuova misura della povertà energetica delle famiglie. *Bank Italy Occas. Pap.* **2017**, *404*. [CrossRef]

38. Climate-KIK. Municipality-led Circular, Economy Case Studies. 2018. Available online: https://www.climate-kic.org/wp-content/uploads/2019/01/Circular-Cities.pdf (accessed on 15 January 2020).
39. Interreg Europe. Renewable Energy Community. A Policy Brief from the Policy Learning Platform on Low-Carbon Economy. 2018. Available online: https://www.climate-kic.org/wp-content/uploads/2019/01/Circular-Cities.pdf (accessed on 15 January 2020).
40. Orientations towards the First Strategic Plan for Horizon Europe. Available online: https://ec.europa.eu/info/sites/info/files/research_and_innovation/strategy_on_research_and_innovation/documents/ec_rtd_orientations-he-strategic-plan_122019.pdf (accessed on 15 January 2020).

MDPI

St. Alban-Anlage 66

4052 Basel

Switzerland

Tel. +41 61 683 77 34

Fax +41 61 302 89 18

www.mdpi.com

Energies Editorial Office

E-mail: energies@mdpi.com

www.mdpi.com/journal/energies